D1272888

Chicago Public Library

REFERENCE

Form 178 rev. 11-00

The Observing Guide to the Messier Marathon

The Messier Catalogue is a list of 110 galaxies, star clusters and nebulae, and includes many of the brightest and best-known objects in the sky. Amateur astronomers can challenge their abilities by attempting to find all of Messier's Objects in one night, and thus complete the Messier Marathon. This book contains over ninety easy-to-use star maps to guide the observer from one object to the next, and provides tips for a successful night of observing. The book also tells the story of the eighteenth century astronomer, Charles Messier, and how he came to compile his extensive Catalogue. This complete guide to the Messier Marathon will help the amateur astronomer to observe the Messier Objects throughout the year, using a telescope or even a pair of binoculars.

DON MACHHOLZ is an amateur astronomer living in Colfax, California. He is an active comet hunter, and has discovered nine comets that now bear his name. He was the Comets Recorder for the Association of Lunar and Planetary Recorders for twelve years. A keen writer, Don was the author of a monthly astronomy column for twenty-two years, and has also written numerous astronomy articles for local newspapers and radio stations. This is his fourth astronomy book.

The Observing Guide to the Messier Marathon

A Handbook and Atlas

Don Machholz

CAMBRIDGE
UNIVERSITY PRESS

PUBLISHED BY THE PRESS SYNDICATE OF THE UNIVERSITY OF CAMBRIDGE
The Pitt Building, Trumpington Street, Cambridge, United Kingdom

CAMBRIDGE UNIVERSITY PRESS
The Edinburgh Building, Cambridge CB2 2RU, UK
40 West 20th Street, New York, NY 10011–4211, USA
477 Williamstown Road, Port Melbourne, VIC 3207, Australia
Ruiz de Alarcón 13, 28014 Madrid, Spain
Dock House, The Waterfront, Cape Town 8001, South Africa

http://www.cambridge.org

First published 2002

Printed in the United Kingdom at the University Press, Cambridge

Typeface Quadraat 12/15 pt   System QuarkXPress™   [SE]

A catalogue record for this book is available from the British Library

Library of Congress Cataloging in Publication data

Machholz, Don, 1952–
     The observing guide to the Messier marathon / Don Machholz.
         p.   cm.
     Includes bibliographical references and index.
     ISBN 0 521 80386 1
         1. Astronomy – Observers' manuals.   2. Astronomy – Charts, diagrams, etc.
     3. Galaxies – Charts, diagrams, etc.   4. Stars – Clusters – Charts, diagrams, etc.
     5. Nebulae – Charts, diagrams, etc.   6. Messier, Charles. Catalogue des nâbuleuses et
     amas d'âtoiles.   I. Title.

     QB64.M234   2002
     522–dc21   2002071562

ISBN 0 521 80386 1 hardback

Contents

Part 2 Atlas

Tables

Preface

Imagine yourself standing next to your telescope at evening twilight. It is late March. The sky is clear, the wind still. It is the night of the Messier Marathon. Tonight you will have the opportunity to locate and observe 110 galaxies, star clusters and nebulae cataloged 200 years ago by a French astronomer named Charles Messier. This basic list contains some of the best astronomical objects ever seen. Most amateur astronomers don't bother finding all of them in a lifetime; you are going to marathon through the whole list in one night.

You begin by working your way upward from the western horizon. After these galaxies you enter the open clusters and nebulae of the winter Milky Way. The variety of these wonders is astonishing.

Time passes quickly. You have been marathoning at a leisurely pace for nearly two hours now. Already you are examining the galaxies high in the sky near the Big Dipper. Next comes the area you have feared the most – the Virgo Galaxies. You set out on it using your trusty star chart. In twenty minutes you have picked up seventeen more Messier Objects. 'This is easy,' you think.

It is now 10:30 PM. You have seen sixty-six of the 110 Messier Objects. You're ahead of schedule. You can now sleep, view other objects, sketch a planet, photograph the sky, watch for meteors, locate Pluto, or visit the other marathoners.

It is now 1:30 AM. The winter Milky Way has set in the west; the summer Milky Way has risen in the east. You pick up some of the most beautiful clusters and nebulae you have ever seen. You swing through Scorpio, then make another sweep down the Milky Way, stopping at each Messier Object in your path.

You are enjoying yourself so much that you are a bit startled when you realize that there is only forty-five minutes until morning twilight. You have seven more Objects to find. This would normally not be difficult, but there is little room for error. You pick up M55, M75, M15, M2, M72 and M73. Twilight approaches. You set your scope on the location for M30, your last Messier Object. There is nothing left to do but stare through your eyepiece. Finally, you see it. In the span of nine hours you have observed the complete Messier Catalogue.

As you put away your telescope, your memory plays back images of everything you have seen. Never before have you observed so many objects in so little time. Every Messier Object remains fresh in your mind.

It's been a good night. You've gained a lot of experience and confidence during the past few hours. You'll be back again next year.

Let's re-design the scene. You miss the first two Messier Objects. Your neighbor burns leaves, engulfing you in thick smoke. Or the Sheriff arrives and tells your group that you must either keep the dirt road clear of cars, or you'll all have to leave. Or your spouse calls you in to put the kids to bed during your second hour. The wind whips up by 10 PM, blowing over your camera and breaking the filter. The Virgo Group takes two hours. Your flashlight batteries begin to die. The fog rolls in one hour before dawn. Do these things happen? Yes. Every one of them has happened to me while marathoning. These same things could happen to you, too. But in every situation I've learned something, grown, and became a better observer. And so will you, no matter how many Messier Objects you see or don't see. After a Marathon I see a lot of tired and happy people, some of the happiest are those who have seen fewer than one hundred Objects – many of these Objects they have never seen before.

It has been more than two decades since I suggested a 'Messier Marathon' to the San Jose Astronomical Association; I've seen it continue to grow as astronomical clubs and individuals around the world take part in this challenge to locate and observe all 110 of the clusters, nebulae and galaxies contained in a catalog compiled by French comet hunter Charles Messier nearly 200 years ago. This book is designed to help the observer to find, in one night, as many Messier Objects as one's experience, instrument, weather, latitude, and ability will allow.

Acknowledgments

When Charles Messier compiled his list of 110 galaxies, star clusters and nebulae over 200 years ago, little did he know that an amateur astronomer would one day write a book to describe how to find the whole batch in one night. This book would not be possible without the efforts of others, and I wish to thank them publicly at this time. Charles Messier wrote his Catalogue. I thank him for including all the wonderful galaxies, nebulae and clusters (even M30).

I thank those who started the Messier Marathon. This includes Tom Reiland, Tom Hoffelder and Ed Flynn. The late Walter Scott Houston helped bring the idea to a universe-full of amateur astronomers.

So many others have helped the Messier Marathon to expand and grow. A.J. Crayon of Arizona has been one of the leaders in developing the Marathon into a spring ritual in the Arizona Desert. Many others have also introduced this idea to their local astronomy clubs.

Two valuable astronomy software programs helped to simplify my making of the sky maps. The software Deep Space 3D™ was used for some of the star maps. THE SKY™ software by Software Bisque was used for others.

When I wanted to illustrate this book I turned to dozens of web sites of beautiful images of the heavens. Deciding which to use was difficult because there were so many to choose from. Thanks to all of you who post these images for us to view.

It was Simon Mitton of Cambridge University Press who urged me to expand upon my writings and maps and make them accessible to a world-wide audience. Compiling this manuscript was not easy. I needed help with my grammar and spelling. Rich Page, a friend, fellow amateur astronomer and a human spellchecker, looked through my writings and offered many suggestions. Most of the mistakes I've made while writing this book will never make it to print because of Rich Page.

A renowned Messier Marathon expert, Hartmut Frommert, reviewed these pages, making suggestions where needed. Hartmut's web site has been most valuable to me as I researched this book.

Even with all the help in writing this book, I still remain responsible for the errors. It sounds unfair, but that is just the way it is.

It has not been easy for my family as I've had to share some of their time with this book. I wish to thank Laura, Matt and Mark for their patient support during this past year.

Don Machholz
August 2001
Colfax, California

Part 1
Handbook

1 Charles Messier

Charles Messier lived and worked during a pivotal point in visual astronomical history. He was one of the first comet hunters, discovering new comets over a span of four decades, and recording nearly every observable comet during his career.

His comet hunting resulted in an extensive knowledge of the night sky, enabling him to organize a catalog of galaxies, clusters and nebulae. This list of heavenly wonders, known as the Messier Catalogue, has become one of the most popular lists of its kind. It includes many of the brightest and best-known objects in the night sky. Yet the 110 marvels are few enough that even the beginning amateur astronomer of today can find them all, or nearly all, of them in one night.

Born on June 26, 1730, in Lorraine, France, Charles was the tenth of twelve children.[1] His father died when he was eleven. Three years later, in early 1744, the young Charles observed the brilliant multi-tailed comet of 1744. A month after his eighteenth birthday, in July 1748, he observed an annular solar eclipse from his home town. In October 1751 he went to Paris in search of a new life. His skill in penmanship and drafting landed him employment as a record keeper at a small observatory at the Hotel de Cluny. One of his first tasks was copying maps of the Great Wall of China and of the City of

Figure 1.1

Charles Messier, 1771. Courtesy of
Dr Owen Gingerich.

what was then called Pekin.[2] Other duties included surveying, making maps of the local area, recording sunspots and compiling meteorological data. His employer, Joseph Nicholas Delisle, gave Messier the position of 'Depot Clerk of the Navy'.

As time went on, Messier was trained to use various telescopes at the observatory to obtain and record exact positions of heavenly bodies. His first documented observation was of the planet Mercury's transit of the sun on May 6, 1753. By 1757 he was searching for a comet famously predicted to return by Edmond Halley. His search was based on Delisle's calculations of the comet's likely position, and was carried out with a 1.5 meter focal length reflector. The diameter of the telescope's mirror was about 0.2 meter, but, being made of metal, it reflected little light and gave poor images.

It was during this period, in 1758, that Messier conceived of his now famous celestial catalog. While tracking yet another comet of that year, Messier noted a strong resemblance between this comet and a nearby nebula. This object, now known as the Crab Nebula, had been discovered twenty-seven years earlier by John Bevis. It occurred to Messier that a catalog giving positions and descriptions of such comet look-alikes would help prevent confusing them with the real thing.

Meanwhile, Messier continued his quest for Halley's Comet, not knowing at the time that Delisle's calculations were flawed and were misdirecting his search. On Christmas night, 1758, an amateur astronomer in Germany, Johann Palitzch, was the first to find Halley's Comet. Messier would find it nearly four weeks later, on January 21, 1759. News of Messier's find was withheld by Delisle until April 1, long after the comet faded into the evening sky and re-appeared in the morning sky. It is not known why Delisle failed to publish Messier's observation in a timely manner, but the practical result was general skepticism of Messier's claim to have found the comet ten weeks earlier. Messier later wrote that the delay in the announcement was one of the biggest disappointments in his life.

Johann Palitzch found no more comets, but Messier continued searching until, by 1801, he had discovered or co-discovered twenty comets. These comet discoveries brought Messier fame and distinction, along with allowing him the pleasure of viewing the night sky through a telescope.

Messier used more than a dozen telescopes during his observing

Table 1.1. **Comets discovered by and credited to Charles Messier.**
The comet designation and comet name are followed by the discovery date, with
the month followed by the day and year. The position is in 2000 coordinates and
indicates the position of the comet at discovery. The magnitude is the brightness
of the comet when found. Elongation is the number of degrees the comet was
from the sun as seen from the Earth. Next we see if the comet was found in the
morning or evening sky. The instrument indicated is either a telescope (T) or the
unaided eye (U).

Comet	Comet name	Disc. date	RA	Decl.	El.	Mag.	Sky	Inst.
C/1760 B1	Messier	01/26/1760	10:56	−15.8	133	5.5	M	T
C/1763 S1	Messier	09/29/1763	16:28	−6.0	59	5.0	E	T
C/1764 A1	Messier	01/04/1764	15:47	+57.7	91	3.0	M	U
C/1766 E1	Messier	03/09/1766	1:23	+16.2	34	6.0	E	T
C/1769 P1	Messier	08/09/1769	2:27	+13.0	101	5.5	M	T
D/1770 L1	Lexell	06/15/1770	18:25	−16.6	169	7.0	E	T
C/1771 G1	Messier	04/02/1771	2:48	+21.3	31	4.5	E	U
C/1773 T1	Messier	10/13/1773	10:27	+5.5	47	4.5	M	T
C/1780 U2	Messier	10/27/1780	11:49	+12.9	46	7.0	M	T
C/1785 A1	Messier–Mechain	01/08/1785	2:20	+5.3	103	6.5	E	T
C/1788 W1	Messier	11/26/1788	11:19	+46.3	96	6.0	M	T
C/1793 S2	Messier	09/28/1793	16:21	+13.9	60	6.0	E	T
C/1798 G1	Messier	04/13/1798	3:30	+24.2	31	6.0	E	T

career. To search for Halley's Comet in 1757–9 he used a reflector
with a mirror of about 0.2 meter diameter. He had access to other
instruments in the observatory, including a refractor with a lens of
about 10 cm, a focal length of 1.1 meter, and a magnification of
120 ×. This refractor seemed to be his favorite instrument, and was
used for much of his observing.

Table 1.1 lists the thirteen comets that were discovered first by
Charles Messier.[3] One, Comet Lexell, was discovered by Messier, but
carries the name of the orbit calculator, Anders Lexell. This
particular comet had a short orbital period until it passed too close
to Jupiter in 1779. This changed the orbit, and the comet is now lost.

Table 1.2 includes seven additional comets found by Messier
shortly after other discoverers found them.[4] Under present
regulations, a comet can carry the names of as many as three
discoverers if their verified independent discovery is made shortly
after the original find. According to these rules it seems unlikely that
any of these comets would also bear Messier's name. In four cases

Table 1.2. **Comets independently discovered by Charles Messier.**
The comet designation and comet name are followed by Messier's discovery date.
The position, magnitude and elongation are for the time Messier found the
comet. 'Sky' indicates morning or evening sky, and 'Late' denotes the number of
days Messier's find followed the original discovery.

Comet	Comet name	Disc. date	RA	Decl.	Mag.	El.	Sky	Late
C/1758 K1	De la Nux	08/14/1758	5:40	+28.5	7.0	60	M	80
P/1758 Y1	Halley	01/21/1759	23:43	+3.0	3.0	53	E	27
C/1760 A1	Great Comet	01/08/1760	7:57	−17.6	2.0	140	M	1
D/1766 G1	Helfenzrieder	04/08/1766	2:50	+26.0	2.5	27	E	7
C/1771 A1	Great Comet	01/10/1771	8:41	+3.2	5.0	156	M	1
C/1779 A1	Bode	01/19/1779	19:27	+29.8	5.0	51	E	13
C/1801 N1	Pons	07/12/1801	6:45	+72.3	6.5	51	M	1

too much time had elapsed. In two more there were so many nearly simultaneous discoveries that assigning credit to a single observer was impossible. In such cases, the comets, appearing quite bright and spectacular, were called 'Great Comets'. In the final case there were several independent discoveries on the same night as Messier's, one night after Pons found this, his first, comet.

A look at Charles Messier's comet discoveries can give us an insight into his comet-hunting activities. It is interesting to note that seven of his twenty finds were made during January. This was despite the fact that in Paris the cloudiest months are December and January; and January has the greatest number of days of precipitation (twenty) and the coldest temperatures.[5] He seemed to sweep equally the morning and evening sky, with the morning comets found at a greater elongation than those found in the evening sky. Whether this was caused by his sweeping (or sleeping!) habits, or from horizon height differences, we do not know. He was also not afraid to sweep through the Milky Way, where many stars and nebulae tend to mask new comets; nor did he shy away from areas containing galaxies. From his latitude of +48.8 degrees, he searched from the North Pole to roughly −20 degree declination.

In Table 1.3 we take a look at the significance that Messier played in the field of comet hunting during his lifetime. Most of his comets were found in his first fifteen years of searching, while nearly half of the objects listed in his Catalogue were recorded sometime later, between 1778 and 1781 (Table 1.4).

Table 1.3 **Comets discovered during each five-year interval, 1758-1802.**
The total number of comets discovered is given for each five-year period. The
number discovered by Messier also includes his independent finds. The number
found by others includes the seven comets which Messier found independently
that do not bear his name, plus the one bearing his name along with Mechain's
(1785 A1).

Years	# Found	# By Messier	# By others	Other discoverers
1758–1762	5	4	4	Klinkenberg
1763–1767	4	4	1	Helfenzrieder
1768–1772	5	4	1	Montaigne
1773–1777	2	1	1	Montaigne
1778–1782	5	2	4	Bode, Mechain
1783–1787	7	1	7	Mechain, C. Herschel
1788–1792	6	1	5	C. Herschel
1793–1797	6	1	5	C. Herschel
1798–1802	6	2	5	Mechain, Pons
Total	46	20	33	

Charles Messier received many awards during his lifetime. In
January 1763 he barely missed being elected to the French Academie
Royale des Sciences. On December 6, 1764, he became a foreign
member of the Royal Society of London. He was also elected
member of the Russian Academy of Sciences. In 1769 he earned
membership in the Berlin Academy of Sciences. In 1771, he was
finally elected to the Paris Academy of Sciences. By then he was also
Astronomer of the Navy, Director of the Observatory at Cluny, and a
member of the Royal Academy of Sciences. From 1785 to 1790 he
held the post of editor of the French journal *Connaissance des Temps*. In
1806 he received the Order of the Cross from Napoleon. These
honors were all in recognition of his careful comet hunting, comet
observing, and his cataloging of nebulae.

On November 26, 1770, Charles Messier married Marie-
Francoise de Vermauchampt. They had met shortly after Messier
arrived in Paris. On March 15, 1772, she gave birth to a son (Antoine-
Charles), who died at the age of eleven days. Messier's wife died
three days earlier, on March 23, 1772.

On November 6, 1781, Messier suffered an 8 meter fall into an ice
cellar. This occurred during the daytime, not at night as some have
reported. Severe injuries resulted, and Messier was sidelined for

Figure 1.2

Comet Hale-Bopp, imaged on March 19, 1997, by Don Machholz.

nearly a year. His first return to the observatory was to observe the transit of the sun by the planet Mercury on November 9, 1782. Perhaps this transit brought back memories of his first official observation, the transit of Mercury twenty-nine years earlier.

During the last decade of the eighteenth century, France was in political turmoil. Science became less important, and those who were politically on the wrong side of the establishment suffered. Messier did less and less observing.

In later years Messier lived by himself, then with his sister and his brother. After they passed away, he lived with a widowed niece, and in 1815 he suffered a stroke. Two years later he contacted dropsy, and after an illness of only a few days, he died on the night of April 11/12, 1817, at the age of eighty-seven.

Table 1.4 **Charles Messier timeline.**

For each year, we see the comets discovered by Messier and the Messier Objects he observed. Then we discover the other activities going on in Messier's life and in the lives of those around him.

Year	Comets found	Obj. found	Other activities
1756			
1757		M32	Begins search for Halley's Comet
1758	1758 K1	M1	Palitzch finds Halley's Comet
1759	Halley's		
1760	1760 A1, 1760 B1	M2	Delisle retires
1761			Observes transit of Venus
1762			Lacallie dies
1763	1763 S1		Almost elected to Academy of Science
1764	1764 A1	M3–M40	
1765		M41	Made Member of Royal Society of London
1766	1766 E1, 1766 G1		
1767			Ocean trip, away from Paris May–Aug.
1768			
1769	1769 P1	M42–M45	First Catalogue written
1770	1770 L1		Married; elected to Academy of Science
1771	1770 A1, 1771 G1	M46–M49, M62	First Catalogue published, M1–45
1772		M50	Wife and son die
1773	1773 T1	M110	
1774		M51, M52	Introduced to Pierre Mechain
1775			
1776			
1777		M53	Messier sees 'specks' crossing sun
1778		M54, M55	
1779	1779 A1	M56–M63	
1780	1780 U2	M64–M79	Second Catalogue published, M1–68
1781		M80–M106, M108–M109	Final Catalogue published; injured in fall; Uranus discovery
1782		M107	Observes transit of Mercury
1783			Herschel begins listing objects
1784			
1785	1785 A1		
1786			Herschel's first catalog published
1787			
1788	1788 W1		

Table 1.4 (cont.) **Charles Messier timeline.**

For each year, we see the comets discovered by Messier and the Messier Objects he observed. Then we discover the other activities going on in Messier's life and in the lives of those around him.

Year	Comets found	Obj. found	Other activities
1789			The Storming of the Bastille
1790			
1791			
1792			
1793	1793 S2		'Year of Terror' in France
1794			
1795			
1796			
1797			
1798	1798 G1		
1799			
1800			
1801	1801 N1		First asteroid discovered

Notes

1 Jones, Kenneth Glyn, *Messier's Nebulae and Star Clusters*, First Edition (1968). Published by American Elsevier Publishing Co., NY, p. 377. Contains an extensive study of Messier's life and the Messier Objects.

2 Mallas, John, and Kreimer, Evered, *The Messier Album*, First Edition (1978). Published by The Nimrod Press, Boston, MA, p. 1. Includes a chapter on Charles Messier, entitled 'Messier and His Catalogue', written by Owen Gingerich. It first appeared in the magazine *Sky and Telescope* in Aug./Sept. 1953 and Oct. 1960.

3 Kronk, Gary W., *Comets: A Descriptive Catalogue*, (1984). Published by Enslow Publishers, Box 777, Hillside, NJ 07205, pp. 18ff. Contains interesting descriptions of comets and comet discoveries up to the end of 1981.

4 Marsden, Brian G., and Williams, Gareth V., *Catalogue of Cometary Orbits*, Tenth Edition (1995). Published by the Smithsonian Astrophysical Observatory, 60 Garden St, Cambridge, MA, 02138. An official listing of all known comets with orbital elements and official names.

5 Sperling, Bert, *Best places* at www.BestPlaces.net. Contains climate and weather for hundreds of locations, including Paris, France.

2 The Messier Catalogue

Lists of non-stellar heavenly bodies – galaxies, clusters and nebulae – were common in Messier's time. Ptolemy compiled one of the earliest such lists in the second century AD.[1] Tycho Brahe had published a list of six nebulae in 1601, as did Edmond Halley in 1715. Abbe Nicholas-Louis de la Caille (Lacaille) produced a tabulation of forty-two objects in the southern sky in 1755, and John Bode published seventy-five objects in 1777.[2]

Perhaps no one was in a better position to compile such a catalog than comet hunter Charles Messier. He had both the means and a motive. Countless nights under the sky sharpened his knowledge of the location and appearance of the objects. This was augmented by his mapping skills.

Messier's main motive for assembling his Catalogue seems to be best summed up in his memoir in the journal *Connaissance des Temps* for 1801. In it he wrote:[3]

> What caused me to undertake the catalogue was the nebula I discovered above the southern horn of Taurus on September 12, 1758, whilst observing the comet of that year. This nebula had such a resemblance to a comet in its form and brightness that I endeavored to find others so that astronomers would not confuse these same nebulae with comets just beginning to appear. I observed further with suitable refractors for the discovery of comets and this is the purpose I had in mind in forming the catalogue.

Figure 2.1

M1, the Crab Nebula. The first object in Messier's Catalogue, the 'Crab Nebula', is now also known as 'Messier 1' or 'M1'. North is up. Field of view: 30 × 20 arcminutes.

Image courtesy of Russ Dickman.

However, the Catalogue was slow in coming. It was two years before Messier recorded his second object, a globular cluster in Aquarius. It had been discovered in 1746 by Jean-Dominique Maraldi. Three more years passed before Messier's third object, a globular cluster in Canes Venatici, was logged. This was his first 'discovery'; that is, this cluster had not been recorded before Messier saw it (Figure 2.2).

Following his find of M3 on May 3, 1764, Messier recorded the positions and descriptions of the next thirty-seven objects within the next seven months. Why so many in so little time? There seem to be three reasons for this. First, it seems probable that Messier mapped clusters and nebulae that he encountered while comet hunting. (He did much of it between 1758 and 1764, finding five comets.) He later returned to each non-comet object to measure its position and write a description of each.[4] Secondly, at about the same time that Messier backtracked through the objects he had discovered, he decided it would be a good idea to examine objects in catalogs compiled by

Figure 2.2

M3: a globular cluster in Canes Venatici. Messier recorded this on May 3, 1764, as his first actual Messier Object 'discovery'. North is up. Field of view: 69 × 62 arcminutes.
Image courtesy of Scott Tucker.

others: Hevelius, Halley, Maraldi, de Cheseaux, LeGentil and Lacaille. He included them in his Catalogue if they met the criterion of being nebulous, and placed them on a separate list of missing objects if they could not be found. Thirdly, his nomination to the Academy of Sciences in 1764, his election to the Royal Society of London in 1765, and his discovery of the bright, periodic comet of 1769 propelled him to the forefront of European astronomy. This notoriety no doubt encouraged Messier in all his astronomical pursuits, including his nebulous object catalog.

Shortly before his fortieth birthday in 1770, Messier had completed the first part of his Catalogue. This included the forty objects he had compiled up to the end of 1764, plus five additional easily seen objects that had been discovered by other observers. Of the total of forty-five objects, seventeen were original Messier discoveries. This first Catalogue appeared in the 1771 edition of the *Memoirs of the Academy of Sciences*.[5]

The tabulation for the final Catalogue began in 1771 and lasted ten years. These objects average 1.8 magnitudes fainter than those in the first Catalogue. Twenty-two of these fifty-eight were first found by Pierre Mechain, Messier's younger friend and fellow comet hunter.

A few objects in the final Catalogue had been seen in 1751 in the Southern Hemisphere by Lacaille, and re-observed by Messier. M56 to M61 were recorded while Messier was following the Comet of 1779; Messier discovered the comet and M56 on the same night. Mechain's discovery of M85 on March 4, 1781, led Messier into this area, where he found an additional seven galaxies in a later examination of the area. The globular cluster M80 was recorded by Messier on the morning of January 4, 1781, when Mars, Jupiter, M80, Venus, Saturn and Mercury were all in a line spanning less than 40 degrees. The complete final Catalogue, numbering Messier Objects 1 to 103, was printed in the *Connaissance des Temps* for 1784, published in 1781.[6]

What Messier called his 'observation date' for each object seems to be the time he measured its position, made a drawing of it, and prepared it for entry into his Catalogue. He claimed to have made a drawing of each object.[7] After his full Catalogue was released, Messier continued to find nebulae and had plans to publish these too. This list surely contained nebulous objects that Messier is now criticized for having missed. In his memoirs appearing in the *Connaissance des Temps* for 1801, he wrote:[8]

> Since the publication of my catalogue I have observed still others: I will publish them in the future in the order of right ascension for the purpose of making them more easy to recognise and for those searching for comets to have less uncertainty.

Such a catalog was never made public. By this time (1786), William Herschel, using a much larger telescope, had published his first list of 1000 new nebulae and clusters. Out of respect to Messier, Herschel did not include the Messier Objects in his own catalog.

As perhaps the first telescopic comet hunter, Messier realized that there are many objects in the heavens that can be mistaken for comets. He plotted them as he saw them in the course of his search for comets. This type of activity still goes on today. Over the years I've received lists of known nebulae and clusters compiled by other comet hunters, plus I've assembled my own list of 253 non-Messier Objects. Messier soon realized that, by publishing his list, he could not only alert other comet hunters of their existence, but could achieve greater status among his peers as well. After his final

Figure 2.3

M24: a star cloud in Sagittarius. North is up. Field of view: 4 by 4 degrees. Image courtesy of Russ Dickman; labeling by the author.

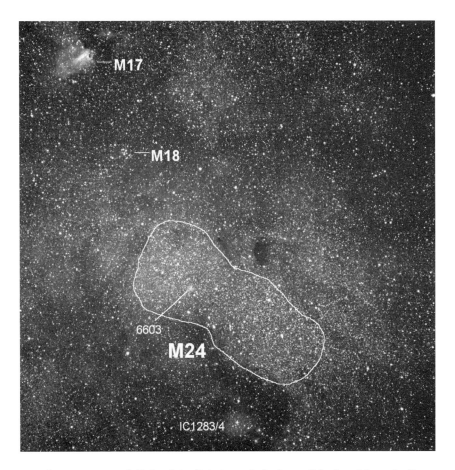

Catalogue was published (1781), he might have felt that Herschel's extensive list made a further catalog unnecessary.

Defining the 'nebulous' objects

While Charles Messier's Catalogue was the best and most accurate produced up until 1786, errors and misunderstandings have existed ever since it was compiled. Here we examine those objects that have been the source of this confusion. For each object, the 'M' preceding the number stands for Messier.

M24

M24 is a star cloud in the Milky Way. Even though there is a small open cluster (NGC 6603) in the area, the cluster is only one part of the nebulous patch, which is about 1 by 2.5 degrees. The large size and low surface brightness of this star cloud make it one of the hardest Messier Objects to observe. Its beauty and position amid the Milky Way makes it one of the most photographed areas.

M40

M40 is a double star in the Big Dipper. This is one of the objects Messier checked out, having learned of it from a catalog compiled by Hevelius. The stars are magnitude 9.0 and 9.3 and are 49 arcseconds apart. Messier did not see any nebulosity associated with them but included it in his list anyway.

The following two objects were recorded by Charles Messier on the same night: February 19, 1771. At that time he plotted four objects, M46 to M49. However, for M47 and M48, no object appears in his indicated position.

M47

It now seems certain that M47 is the open cluster NGC 2422. Apparently Messier made an error in the signs indicating the offset of this cluster from his comparison star. Credit for this identification goes to both Oswald Thomas (1934) and T.F. Morris (1959).

M48

This is the open cluster NGC 2548. Messier's position is at the same right ascension as NGC 2548, but apparently the declination is off by 5 degrees. T.F. Morris suggests that this is NGC 2548, the only bright cluster in the area.

M73

M73 is a small cluster of four stars. Some historians have said that this is too insignificant to be a true Messier Object, but a cluster of four stars is still a cluster. Messier knew what he was looking at, describing it as 'a cluster of three or four small stars'.

M91

This was the last of nine galaxies observed by Messier on the night of March 18, 1781. Nothing appears in his original position. We are now rather certain that this is the galaxy NGC 4548.

For many years it was thought that this was a comet and that Messier did not detect its motion and therefore its true identity.

Chart 2.1. M91

The 'X' denotes Messier's position for M91. Field of view: 4.0 by 4.7 degrees. North is up. Stars to magnitude 14.0. THE SKY™ software by Software Bisque.

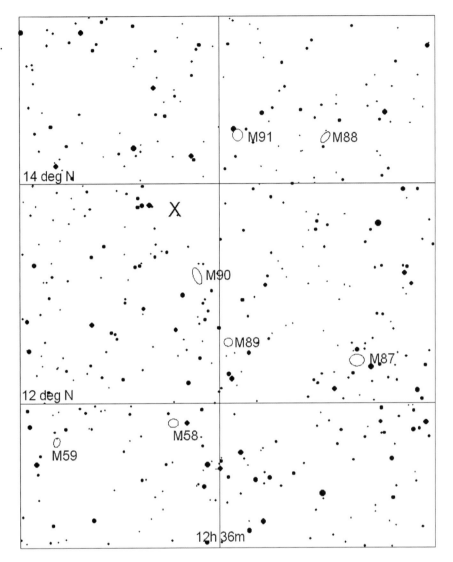

Owen Gingerich suggested in 1960[9] that Messier made a sign error in his offset and thought M91 was above, not below, M90. Therefore, he concluded, perhaps Messier was actually re-observing M58. The problem is that Messier states that M91 is fainter than M90, whereas M58 is brighter.

In 1969, William Williams,[10] an amateur astronomer from Texas, suggested that M91 is NGC 4548. The position agrees very well if Messier used M89 as a reference for offset when he actually thought he was using M58. This seems to be the final word in the matter, and M91 is now taken to be NGC 4548.

An interesting aside is that Messier's erroneous plot of M91 appears on his map of the comet of 1779.[11] The map was completed in 1779, but Messier added and numbered the objects later. M91 is indicated by an '11'. This suggests that Messier plotted his objects from their coordinates, rather than while at the telescope.

M102

For the Messier Marathon, M102 is taken to be the galaxy NGC 5866. Perhaps no object has caused as much controversy as M102, and arguments can be made for two sides of the issue.

This object was added to the Catalogue just before publication in 1781. Therefore, Messier did not have time to check the position or appearance of this object, and perhaps he did not even observe it. Indeed, it was his assistant, Pierre Mechain, who discovered it and described it as 'a very faint nebula situated between Omicron Bootis and Iota Draconis: near to it is a 6 mag. star'. In addition, Mechain probably made a rough estimate of its coordinates, but did not publish them, waiting, instead, for Messier to observe the object, write a description, and measure accurate coordinates. Apparently Messier did observe it and generate a rough position for it, as we will see later.

Already we find an error in the description. Omicron (*o*) Bootis is at roughly right ascension 14hr 42min, declination +17 degrees. Iota Draconis is at right ascension 15hr 25min, declination +59 degrees. This is a difference of more than 40 degrees, and it is unlikely that Mechain would try to locate a galaxy by such divergent guideposts. One of these two stars has been misidentified. If Omicron Bootis is correct, is there any star nearby that Mechain might have mistaken for Iota Draconis? The constellation Draco stays north of declination +48 degrees, over 30 degrees from Omicron Bootis. Moreover, no bright stars named Iota are near Omicron Bootis, which itself is nestled among other stars better located for reference.

More than likely, Mechain meant Theta (*θ*) Bootis. Because of the similarity of the two symbols, we will make the assumption that this error was made, and on this we will base the remainder of this discussion. There are three possible causes for such an error: (1) he miswrote the symbol, (2) he misread the symbol on the map, or (3) the map contained the incorrect symbol. In any event, these two stars are 11 degrees apart, and in between them lie several galaxies,

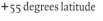
+55 degrees latitude

Chart 2.2. M101 and M102

The 'X' denotes Messier's position for M102. North is up. THE SKY™ software by Software Bisque.

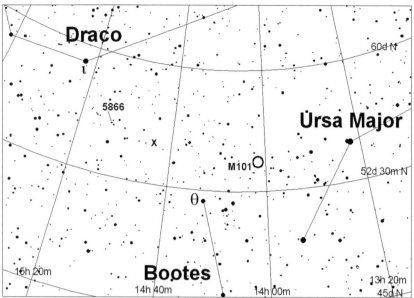

the brightest of which is NGC 5866. Moreover, a star of magnitude 5.2 (SAO 029402) lies 1.3 degrees south of the galaxy, while a 6.8 magnitude star lies about 0.4 degrees away. The galaxy is small and has a high surface brightness; it is certainly easier to see than M76, and is similar in appearance to M77 in a small telescope. Since NGC 5866 would have been easily visible to Mechain, is in the correct position, and has some stars nearby of sixth magnitude, why hasn't everyone generally recognized it as M102?

Shortly after the Catalogue was published, Mechain wrote a letter to J. Bernoulli in Berlin, in which he claims that M102 is actually M101. He states:[12]

> On page 267 of the *Connaissance des Temps* for 1784, M. Messier lists under No.102 a nebula which I have discovered between omicron Bootis and iota Draconis: this is nothing but an error. This nebula is the same as the preceding No.101. In the list of my nebulous stars communicated to him M. Messier was confused due to an error in the sky chart.

Unfortunately, this does not completely clear up the matter. As Messier historian Glyn Jones points out, NGC 5866 and M101 have nearly the same declination but differ in almost exactly 1 hour of right ascension. Could Mechain, while plotting M102 by perhaps 1 hour in error, have seen M101 there and assumed that they were one and the same?[13] If this is the case, then maybe he was observing NGC 5866 after all.

It is difficult to 're-construct' the error. From Mechain's description, we do not know whether this was Messier's or Mechain's mistake. Perhaps Mechain observed M101, thought he was seeing a new nebula (even though he had seen M101 only a few months before), and gave an erroneous position to Messier, who plotted it on the map. Or perhaps Mechain gave the correct position to Messier, but Messier misplotted it so he was unable to tell Mechain that he was again observing M101. In either case, in the 1781 Catalogue Messier had hand-written positions for both M101 and M102. But, while the position of M102 doesn't correspond with a known nebula, neither is it superimposed upon M101.[14]

This is further complicated by something I recently discovered while researching this area. Messier's original Catalogue lists an incorrect position for M101. While the declination is accurate to within 3 arcseconds, the right ascension is off by almost exactly 12.0 minutes, which at this declination equals 1.72 degrees. In 1781 coordinates, Messier lists M101 at 13hr 43min 28s, whereas M101 was actually at 13hr 55min 26s (also 1781 coordinates).[15] This brings forth another possibility: that M101 was observed twice, the first time it was plotted incorrectly and labeled M101; the second time it was plotted correctly and initially called M102. Figures 2.4(a) and (b) show M101 and M102 at approximately the same scale.

However, even if M102 is identical to M101, two rather difficult questions remain.

(1)   Where is the sixth magnitude star that is said to be 'nearby'? M101 has several stars north of it, among which are a 5.7 magnitude star that is 1.4 degrees away, and a 6.8 magnitude star that is 0.8 degree away. NGC 5866 is probably a better candidate on this issue.

(2)   Why is the location given as being between a star in Draco and a star in Bootes when it is in Ursa Major? Messier was familiar with the 'Great Bear', mentioning it in reference to M40, M81 and M101. Yet the fact remains that M102 was described as being between Iota Draconis and (Theta) Bootis, a location which is far from M101 and very close to NGC 5866.

One too must wonder why Mechain, in his letter to Bernoulli, repeats the error of using Omicron Bootis? Could this error have led him to realize, while attempting to plot M102, that the exact position

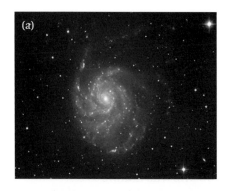

Figure 2.4(a) M101 and (b) M102

Seen at approximately the same scale. Image of M101 courtesy of P. Justis; image of M102 courtesy of Dan Thurs.

would be impossible to find? Or did he spot the juxtaposition of Omicron and Theta but failed to point that out in the letter? Adding to the problem is the fact that Mechain wrote to a German publication to report the mistake and apparently never reported it to the French journal *Connaissance de Temps* that originally printed the Catalogue. This latter would have been easier to do, especially since Messier was an associate editor of the journal from 1785 to 1790. Finally, after its publication in Germany in 1786, the letter 're-surfaced' only twice – in 1877 and in 1917 – before being fully published and recognized in 1947.

It would have solved the problem if the description had read: 'M102: looks something like M101'. Better still would have been: 'M102: nebulae without star, very obscure and pretty large, 6′ or 7′ diameter between the left hand of Bootes and the tail of the Great Bear. Difficult to distinguish when graticule lit', because that is the description Mechain gave to M101.

New information has recently come to light that lends additional credence to M102 being NGC 5866.[16] In Messier's copy of his published Catalogue Messier had hand-written the position for M102. It reads a right ascension (RA) of '14.40' and a declination of '56'. This means an RA of 14hr 40min and a declination of +56 degrees. Precessed to the year 2000, these coordinates become 14hr 46.5min, +55.1 degrees. The position of NGC 5866 is 15hr 06.5min, +55.7 degrees. The galaxy is almost exactly 5 degrees (20 minutes) east of Messier's assigned position. Since we know that Messier's maps had grids 5 degrees apart, could he not have observed the galaxy NGC 5866, but then reduced the right ascension wrongly by reading the wrong grid? When he had plotted M48 he was also off by five degrees, but then in declination. Here, if he misread his map, he was surely seeing NGC 5866.

Until the time that we find all of Messier's and Mechain's original notes on the object, M102 will remain 'up in the air'. Yes, it may have been a duplicate of M101, but on the night of the Messier Marathon it would still be a good idea to find and observe NGC 5866.[17]

The 'add-on' Messier Objects

Although Messier's original Catalogue contained 103 objects, seven additional objects (six of which are galaxies) have been added to his list. These seem to be justifiable additions and are generally recognized to be part of the Messier Catalogue. Here are a few notes about each.

M104 (NGC 4594): Added in 1921 by Camille Flammarion who found it on Messier's copy of his 1781 Catalogue. He described it as 'a very faint nebula in Virgo'.[18]

M105 (NGC 3379), M106 (NGC 4258) and M107 (NGC 6171): In a letter dated 1783, Mechain included these objects, found by him.

M108 (NGC 3556) and M109 (NGC 3992): Messier observed both of these objects on March 24, 1781, and included them in his description of M97.

M110 (NGC 205): On Messier's map of M31 he labels two other galaxies, one being M32, the other being this object, along with the notation that he discovered it in 1773. This date would have placed it between his logging of M50 (April 5, 1772) and M51 (January 1, 1774).

Putting it all together

A worthy goal for each amateur astronomer is to locate, identify and observe each Messier Object. I did this with a 6 inch reflector between May 30, 1969, when I saw M44 and M67, and May 27, 1970, when I finished up with M61 and M104. It was easy to achieve in one year. In chapter 3 I will explain how this can be done in one night.

Table 2.1 **The Messier Catalogue.**

M #	NGC #	Type	RA 2000	Decl. Equinox	Mag.	Size	Messier obs. date	Notes
The first Catalogue								
1	1952	DN	05h 34.5m	+22° 01′	8.7	7′×4′	9/12/1758	Crab Nebula
2	7089	GL	21h 33.5m	−00° 49′	7.0	8′	9/11/1760	'resembles' M22
3	5272	GL	13h 42.2m	+28° 23′	6.4	8′	5/3/1764	'always very fine'
4	6121	GL	16h 23.6m	−26° 31′	6.4	14′	5/8/1764	'near Antares'
5	5904	GL	15h 18.5m	+02° 05′	6.2	10′	5/23/1764	'contains no star'
6	6405	OC	17h 40.0m	−32° 15′	5.3	30′×20′	5/23/1764	'cluster of fine stars'
7	6475	OC	17h 54.0m	−34° 49′	5.0	32′×26′	5/23/1764	Ptolemy, 120 AD
8	6523	DN	18h 03.7m	−24° 23′	5.2	25′×20′	5/23/1764	Lagoon Nebula
9	6333	GL	17h 19.2m	−18° 31′	7.3	4′	5/28/1764	'it is round'
10	6254	GL	16h 57.2m	−04° 06′	6.7	9′	5/29/1764	'fine and round'
11	6705	OC	18h 51.1m	−06° 16′	6.0	8′	5/30/1764	'with a faint glow'
12	6218	GL	16h 47.2m	−01° 57′	6.8	8′	5/30/1764	'contains no star'
13	6205	GL	16h 41.7m	+36° 28′	6.2	13′	6/1/1764	Hercules Cluster
14	6402	GL	17h 37.6m	−03° 15′	7.7	8′	6/1/1764	'nebula without star'
15	7078	GL	21h 30.0m	+12° 10′	6.8	7′	6/3/1764	'nebula without star'
16	6611	OC	18h 18.9m	−13° 47′	6.8	25′×25′	6/3/1764	Eagle Nebula
17	6618	DN	18h 21.1m	−16° 10′	6.6	15′×8′	6/3/1764	Omega Nebula
18	6613	OC	18h 19.9m	−17° 08′	7.1	8′×6′	6/3/1764	'cluster of small stars'
19	6273	GL	17h 02.6m	−26° 16′	7.4	6′	6/5/1764	'round'
20	6514	DN	18h 02.4m	−23° 02′	6.8	16′×10′	6/5/1764	Trifid Nebula
21	6531	OC	18h 04.7m	−22° 30′	7.1	7′×4′	6/5/1764	'cluster of stars'
22	6656	GL	18h 36.4m	−23° 54′	5.4	17′	6/5/1764	'contains no star'
23	6494	OC	17h 56.9m	−19° 01′	6.2	20′×15′	6/20/1764	'stars close together'
24	—	SC	18h 18.4m	−18° 25′	4.5	80′×35′	6/20/1764	large star cloud
25	—	OC	18h 31.7m	−19° 07′	5.4	35′×30′	6/20/1764	'cluster of small stars'
26	6694	OC	18h 45.2m	−09° 24′	7.8	7′	6/20/1764	'no nebulosity'
27	6853	PL	19h 59.6m	+22° 43′	8.1	7′	7/12/1764	Dumbbell Nebula
28	6626	GL	18h 24.6m	−24° 52′	7.8	5′	7/27/1764	'contains no star'
29	6913	OC	20h 24.0m	+38° 31′	7.4	7′	7/29/1764	'7 or 8 stars'
30	7099	GL	21h 40.4m	−23° 11′	8.1	4′	8/3/1764	'seen with difficulty'
31	224	EG	00h 42.7m	+41° 16′	4.5	145′×20′	8/3/1764	Andromeda Nebula
32	221	EG	00h 42.7m	+40° 52′	8.6	2′	8/3/1764	'round'
33	598	EG	01h 33.8m	+30° 39′	6.2	22′×18′	8/25/1764	'seen with difficulty'
34	1039	OC	02h 42.0m	+42° 47′	6.3	25′×15′	8/25/1764	'cluster of small stars'
35	2168	OC	06h 08.8m	+24° 20′	4.8	25′×18′	8/30/1764	'very small stars'
36	1960	OC	05h 36.1m	+34° 08′	5.8	13′	9/2/1764	'no nebulosity'
37	2099	OC	05h 52.4m	+32° 33′	6.2	18′	9/2/1764	'very crowded'

Table 2.1 (cont.)

M #	NGC #	Type	RA 2000	Decl. Equinox	Mag.	Size	Messier obs. date	Notes
The first Catalogue								
38	1912	OC	05h 28.7m	+35° 50'	6.0	11'	9/25/1764	'this one is rectangular'
39	7092	OC	21h 32.3m	+48° 26'	5.4	30'×20'	10/24/1764	'a cluster of stars'
40	—	DS	12h 22.2m	+58° 05'	8.7	1'	10/24/1764	'two stars very close'
41	2287	OC	06h 47.0m	−20° 45'	5.8	25'	1/16/1765	'cluster of small stars'
42	1976	DN	05h 35.3m	−05° 23'	4.3	40'×30'	3/4/1769	Great Nebula in Orion
43	1982	DN	05h 35.5m	−05° 16'	8.3	5'	3/4/1769	just north of M42
44	2632	OC	08h 40.0m	+20° 00'	3.3	60'×50'	3/4/1769	Beehive Cluster
45	—	OC	03h 47.5m	+24° 07'	2.0	70'×40'	3/4/1769	The Pleiades
The second Catalogue								
46	2437	OC	07h 41.8m	−14° 49'	6.7	18'×15'	2/19/1771	'cluster of small stars'
47	2422	OC	07h 36.6m	−14° 29'	5.6	17'×12'	2/19/1771	'no nebulosity'
48	2548	OC	08h 13.8m	−05° 48'	6.2	40'×35'	2/19/1771	'cluster of small stars'
49	4472	EG	12h 29.8m	+08° 00'	8.6	4'	2/19/1771	'seen with difficulty'
50	2323	OC	07h 03.0m	−08° 21'	6.4	8'×6'	4/5/1772	'cluster of small stars'
51	5194	EG	13h 29.9m	+47° 12'	7.9	14'×11'	1/11/1774	'very faint nebula'
52	7654	OC	23h 24.2m	+61° 36'	7.6	9'×7'	9/7/1774	'with nebulosity'
53	5024	GL	13h 12.9m	+18° 10'	7.4	3'	2/26/1777	'nebula without stars'
54	6715	GL	18h 55.1m	−30° 28'	7.7	3'	7/24/1778	'very faint nebula'
55	6809	GL	19h 40.0m	−30° 57'	6.1	12'	7/24/1778	'contains no stars'
56	6779	GL	19h 16.6m	+30° 11'	8.8	5'	1/23/1779	'having little light'
57	6720	PL	18h 53.6m	+33° 02'	8.8	2'	1/31/1779	'mass of light'
58	4579	EG	12h 37.7m	+11° 49'	9.2	4'	4/15/1779	'very faint nebula'
59	4621	EG	12h 42.0m	+11° 39'	9.4	3'×2'	4/15/1779	'as faint (as M58)'
60	4649	EG	12h 43.7m	+11° 33'	9.2	4'×3'	4/15/1779	'disc. while obs. comet'
61	4303	EG	12h 21.9m	+04° 28'	9.0	5'	5/11/1779	'very faint'
62	6266	GL	17h 01.2m	−30° 07'	7.5	5'	6/4/1779	'very faint nebula'
63	5055	EG	13h 15.8m	+42° 02'	8.7	8'×4'	6/14/1779	'it is faint'
64	4826	EG	12h 56.7m	+21° 41'	8.7	8'×4'	3/1/1780	Blackeye Galaxy
65	3623	EG	11h 18.9m	+13° 06'	8.9	11'×4'	3/1/1780	'very faint'
66	3627	EG	11h 20.2m	+12° 59'	8.6	9'×5'	3/1/1780	'its light very faint'
67	2682	OC	08h 51.4m	+11° 48'	7.6	12'×10'	4/6/1780	'cluster of small stars'
68	4590	GL	12h 39.5m	−26° 45'	8.1	5'	4/9/1780	'it is very faint'
69	6637	GL	18h 31.4m	−32° 21'	7.9	3'	8/31/1780	'very faint'
70	6681	GL	18h 43.2m	−32° 17'	8.4	2'	8/31/1780	'nebula without stars'
71	6838	GL	19h 53.7m	+18° 47'	8.1	7'	10/4/1780	'very faint'
72	6981	GL	20h 53.5m	−12° 32'	9.0	4'	10/4/1780	'as faint' as M71

Table 2.1 (cont.)

M #	NGC #	Type	RA 2000	Decl. Equinox	Mag.	Size	Messier obs. date	Notes
The second Catalogue								
73	6994	GL	20h 59.0m	−12° 38′	9.2	2′	10/4/1780	'3 or 4 small stars'
74	628	EG	01h 36.7m	+15° 47′	9.1	9′	10/18/1780	'nebula without stars'
75	6864	GL	20h 06.1m	+21° 55′	8.7	2′	10/18/1780	'very small stars'
76	650	PL	01h 42.2m	+51° 44′	9.6	3′×2′	10/21/1780	'small and faint'
77	1068	EG	02h 42.7m	−00° 01′	8.9	2′	12/17/1780	'cluster of small stars'
78	2068	DN	05h 46.7m	+00° 04′	8.4	8′	12/17/1780	'much nebulosity'
79	1904	GL	05h 24.2m	−24° 31′	8.1	3′	12/17/1780	'nebula is fine'
80	6093	GL	16h 17.0m	−22° 59′	8.1	3′	1/4/1781	'round, bright centre'
81	3031	EG	09h 55.6m	+69° 04′	7.3	12′×5′	2/9/1781	'is a little oval'
82	3034	EG	09h 55.8m	+69° 42′	8.3	9′×2′	2/9/1781	'faint and elongated'
83	5236	EG	13h 37.0m	−29° 52′	7.9	10′×7′	2/17/1781	'faint, even light'
84	4374	EG	12h 25.1m	+12° 53′	8.8	5′	3/18/1781	'resemble' M59, M60
85	4382	EG	12h 25.4m	+18° 11′	8.9	4′	3/18/1781	'very faint'
86	4406	EG	12h 26.2m	+12° 57′	8.8	5′	3/18/1781	'appears like' M84
87	4486	EG	12h 30.8m	+12° 23′	8.7	5′	3/18/1781	'looks like' M84 and M86
88	4501	EG	12h 32.0m	+14° 25′	9.4	8′×3′	3/18/1781	'faint, like' M58
89	4552	EG	12h 35.7m	+12° 33′	9.1	3′	3/18/1781	'faint and diffuse'
90	4569	EG	12h 36.8m	+13° 10′	9.3	6′×2′	3/18/1781	'as faint as ... M89'
91	4548	EG	12h 35.4m	+14° 30′	9.9	3′	3/18/1781	'fainter than' M90
92	6341	GL	17h 17.1m	+43° 08′	6.9	8′	3/18/1781	'fine, conspicuous'
93	2447	OC	07h 44.6m	−23° 53′	5.8	12′×8′	3/20/1781	'cluster of fine stars'
94	4736	EG	12h 50.9m	+41° 07′	8.4	4′	3/24/1781	'nebula without stars'
95	3351	EG	10h 44.0m	+11° 42′	9.5	7′	3/24/1781	'very faint'
96	3368	EG	10h 46.8m	+11° 49′	9.1	8′×4′	3/24/1781	'less distinct' than M95
97	3587	PL	11h 14.9m	+55° 01′	9.7	6′	3/24/1781	'difficult to see' Owl
98	4192	EG	12h 13.8m	+14° 54′	9.8	8′×3′	4/13/1781	'extremely faint light'
99	4254	EG	12h 18.8m	+14° 25′	9.4	6′	4/13/1781	'very dim light'
100	4321	EG	12h 22.9m	+15° 49′	9.6	7′	4/13/1781	'nebula without star'
101	5457	EG	14h 03.2m	+54° 21′	8.8	14′	3/27/1781	'very obscure'
102	5866	EG	15h 06.5m	+55° 46′	9.6	4′×2′	?	'very faint'
103	581	OC	01h 33.1m	+60° 42′	7.0	8′	?	'cluster of stars'
The add-on objects								
104	4594	EG	12h 40.0m	−11° 37′	8.8	7′×4′	1921	Sombrero Nebula
105	3379	EG	10h 47.9m	+12° 35′	9.1	4′	1947	Near M95 and 96
106	4258	EG	12h 19.0m	+47° 18′	8.8	11′×5′	1947	spiral galaxy
107	6171	GL	16h 32.5m	−13° 03′	8.6	5′	1947	small globular cluster

Table 2.1 (cont.)

M #	NGC #	Type	RA 2000	Decl. Equinox	Mag.	Size	Messier obs. date	Notes
The add-on objects								
108	3556	EG	11h 11.6m	+55° 40′	9.4	7′×2′	1960	near M97
109	3992	EG	11h 57.7m	+53° 22′	9.6	4′	1960	near Gamma Ursa Major
110	205	EG	00h 40.3m	+41° 41′	8.2	10′×4′	1966	near M31 and M32

Notes:

M # = Messier Object number; NGC = New General Catalogue number, if the Messier Object has one; Type: DN = diffuse nebula, GL = globular cluster, OC = open cluster, EG = galaxy, PL = planetary nebula and SC = star cloud, DS = double star; RA and Decl. = right ascension and declination, i.e. its coordinates in the year 2000 equinox; Mag. = magnitude or brightness of the objects, estimated by Don Machholz; Size = apparent angular size, in arcminutes, estimated by Don Machholz; Messier obs. date: the date listed is the first that Messier observed or recorded the object. For the latter objects the date that it was added to the Catalogue is stated; and Notes: some of the comments made by Messier when he published the Catalogue. One can see the limitations of his instruments. Often he could not see stars in many of the globular clusters; at other times he saw nebulae where there were none.

Notes

1   Jones, Kenneth Glynn, *The Search for the Nebulae* (1975). Published by Alpha Academic, available from Neale Watson Academic Publishers, 156 Fifth Ave., New York, NY 10010, pp. 5–6.

2   *Ibid.*, pp. 11, 22, 44, 53.

3   English translation in Jones, Kenneth Glynn, *Messier's Nebulae and Star Clusters*, First Edition (1968). Published by American Elsevier Publishing Co., NY, p. 398. (Second Edition, 1991, p. 365.)

4   For instance, Messier measured the planetary nebula M27 on July 12, 1764, the night of a Full Moon! Most of the objects were measured when they were near opposition (in the sky for nearly the whole night) and when the moon was not in the sky. This was different from his mode of operation for comet hunting. Most of his comet discoveries were made in parts of the sky that were close to the sun.

5   Jones, *Messier's Nebulae, op. cit.*, pp. 384, 388. (Second Edition, pp. 351, 355.)

6   Two other publications of Messier's Catalogue are mentioned by Owen Gingerich in his August 1953 issue of *Sky and Telescope*. John Mallas and Evered Kreimer reproduce this article in the book *The Messier Album*, First Edition (1978). Published by the Nimrod Press, Boston, MA. Gingerich notes that, in 1780, a list of Messier's Objects up to number 68 was published. Then, in 1787, the complete list, errors and all, was published again.

7   Jones, *Messier's Nebulae, op. cit.*, p. 389. (Second Edition, p. 357.)

8   *Ibid.*, p. 398. (Second Edition, p. 365.)

9   'The Missing Messier Objects' by Owen Gingerich. *Sky and Telescope*, October 1960, pp. 196–199.

10 Letter to the Editor by William Williams. *Sky and Telescope*, December 1969, p. 376.

11 This map can be seen in *Sky and Telescope*, October 1960, p. 198.

12 Jones, *Messier's Nebulae, op. cit.*, p. 35. (Second Edition, p. 14.)

13 *Ibid.*

14 Mallas and Kreimer, *The Messier Album, op. cit.*, p. 13.

15 Mechain was known to have published the position for only one other object: M97. Here again his declination was accurate, but the right ascension was off by 0.85 minutes, which equals 0.12 degrees or 7.1 arcminutes.

16 Frommert, Hartmut, 'M102 Controversy'. *Saguaro Astronomy Club Newsletter*, March 1998, Issue #254. Also at http://www.seds.org/messier/.

17 The late Walter Scott Houston believed that M102 was a duplicate observation of M101 (*Sky and Telescope*, June 1970, p. 402, and July 1980,

p. 81), but he apparently was not opposed to the idea of using M102 for the Marathon. He happened to be present on February 4, 1982, when I delivered a talk to the Tamalpais Astronomical Association concerning the Messier Marathon. After my talk, Houston was asked if he had anything to say. His reply was 'Well, if using 5866 for M102 is good enough for Don Machholz, then it's good enough for me.'

18 Jones, *Messier's Nebulae, op. cit.*, p. 35. (Second Edition, p. 15.)

3 The Messier Marathon

As seen from Chart 3.1, the Messier Objects are not evenly distributed throughout the sky. A large concentration of galaxies appears in the constellation Virgo, and across the border in the small constellation Coma Berenices. Another group of Objects, consisting mostly of open and globular star clusters, resides in Sagittarius. Meanwhile, a few areas in the sky contain no Messier Objects, notably near the Great Square of Pegasus, but also between Ursa Major and Gemini.

As the sun travels through the sky each year on the ecliptic (see Chart 3.1), the adjacent portions of the sky are not easily visible from the Earth because they rise and set with the sun. For example, when the sun is 'in' Gemini, as it is during early July, then it is located in front of the stars that form the constellation that we call Gemini. The stars of Gemini rise at the same time as the sun does. And, as even the non-astronomer notices, the sky brightens for at least an hour before sunrise, making observation of these stars increasingly difficult as dawn advances. Therefore the stars that rise shortly before Gemini are invisible as well. At the end of the day, Gemini sets with the sun. This is followed by a lingering twilight that prevents astronomical observation for perhaps another hour. During this hour, a section of the sky following Gemini sinks down

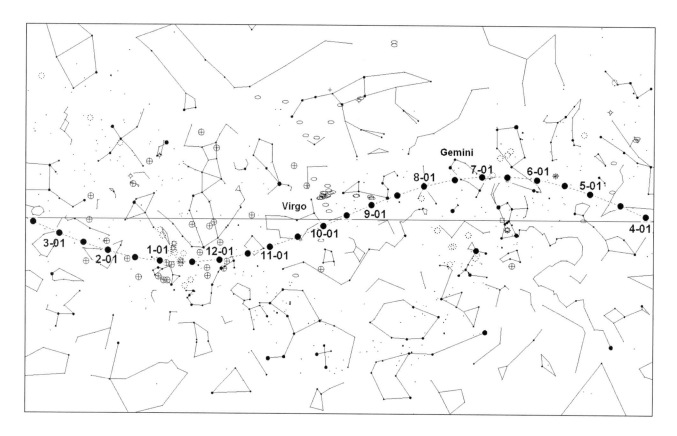

Chart 3.1

The Messier Objects with the position of the sun marked twice each month.
In this chapter, dotted circles indicate open clusters; circles with plus signs indicate globular clusters; ovals represent galaxies; and squares denote nebulae. All maps in this chapter have been produced by the Deep Space 3D™ software program.

below the horizon so that, when the sky finally darkens, this section is not visible.

The fans of Gemini need not despair, however. Besides moving across the sky each day, the sun moves slowly in reference to these background stars. This movement occurs eastward at a rate of about 1 degree a day (or 360 degrees in one year, 365.25 days). This means that a star rises about four minutes earlier each day; and sets four minutes earlier each night. An observer watching the pre-dawn sky each morning in early August would see Gemini rising higher and higher in the eastern sky before the stars would disappear in the morning twilight.

On the other hand, a fan of the constellation Virgo, and all the galaxies associated with it, would see this group setting in the western sky four minutes earlier each evening. Once it disappears in the western sky during early August, it will not be seen again until it rises in the morning sky two months later.

A subtle and often overlooked effect occurs near the spring and fall equinoxes. During spring, the days are getting longer, and the nights are becoming shorter. Evening twilight is perhaps a minute later each night. Therefore, as darkness falls each successive evening, the sky has advanced even more than it usually would. For a few weeks these constellations appear to be on an accelerated advance to the horizon. In the morning sky, twilight occurs earlier each day, meaning the sky doesn't climb as high each successive day as it usually would. Just the opposite happens each fall: the evening sky tends to stall, while the morning constellations advance upward rapidly.

The declination (or 'latitude in the sky') of an object also affects its visibility. For an observer in the northern hemisphere, an area of the sky which is far enough north can be seen both in the northwest after evening twilight and in the northeast before morning twilight. Sky which is too far south will never rise, while sky near the North Pole is above the horizon all night long, every night of the year, and it is said to be 'circumpolar'.

Three key points need to be remembered here:

(1)   we lose areas in the evening sky and gain them in the morning sky;
(2)   during the spring, we lose objects in the evening more quickly, while we gain them in the morning more slowly;
(3)   for northern hemisphere observers, the further north an object is, the less time it will spend hidden by evening and morning twilight.

With this in mind, we now look at the visibility of the Messier Objects throughout the year.

A short history of the Messier Marathon

The visibility of nearly all of the Messier Objects during late March is apparent from Chart 3.1. I realized in the late 1960s that this occurs, but did not begin further investigation until the summer of 1978. In the September 1978 issue of the *San Jose Astronomical Association Newsletter*, I wrote an article entitled 'Messier Marathon', inviting members of the club to join me the following March on Loma Prieta Mountain, my normal comet-hunting site.

Over the next few months, I worked out the observing order, or 'search sequence', based upon the commonly used *Atlas of the Heavens* star atlas. In order to calculate when each object would rise and set, I used a planisphere developed by David Chandler. To determine the date that each object would first become visible in the

Figure 3.1

The author (left) and Walter Scott Houston in February 1982. Courtesy of Laura Machholz.

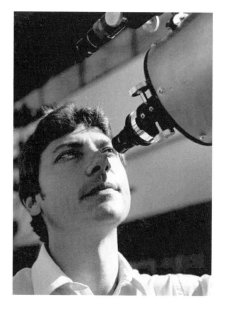

Figure 3.2

Tom Hoffelder at the eyepiece of his telescope. Courtesy of Tom Hoffelder.

morning sky, or last visible in the evening sky, I depended upon my comet-hunting records. Since I recorded every object seen during each session, and I was often sweeping near the horizon at twilight, I had an ideal estimate of what was visible when. Because I had never heard of anyone running a Messier Marathon, I believed that we were the only astronomers working towards one.

Imagine my surprise upon receiving the March 1979 issue of *Sky and Telescope* and reading Walter Scott Houston's (Figure 3.1) 'Deep-Sky Wonders' column describing a Messier Marathon! The article portrayed the Amateur Astronomers of Pittsburgh (Pennsylvania), which started marathoning in 1977.[1] Tom Hoffelder (Figure 3.2), Ed Flynn (Figure 3.3) and Tom Reiland (Figure 3.4) all found a large number of Messier Objects in March/April 1977.

With the use of a descriptive sky map, Walter Scott Houston explained that M30 is the only object not visible during late March of each year. He mentioned that the record for the greatest number of Messier Objects found in one night was 103, held by Tom Reiland of Pittsburgh, while Tom Hoffelder had found 101 Objects.

This article, the first Messier Marathon article ever widely published, resulted in several letters to Walter Scott Houston. A number of groups claimed to have invented the idea. Someone from Spain wrote to say that Houston had 'stolen' the idea; that it had

Figure 3.3

Ed Flynn with his 10 inch reflector in late 1978. Courtesy of Ed Flynn.

Figure 3.4

Tom Reiland, one of the co-founders of the Messier Marathon. Courtesy of Tom Reiland.

been published eleven years earlier.[2] Since then I have heard of an amateur group in Southern California which had carried out informal Messier Marathons since the early 1970s.

For the Pittsburgh Pennsylvania group, the idea originated when Tom Reiland, in May 1975, observed fifty Messier Objects in one night and suggested that an attempt to find a great number of Objects in one night would be an interesting challenge. His first attempt was in April 1976, when he saw eighty-six Objects before he was clouded out. On March 24–25, 1977, Ed Flynn saw ninety-seven Objects from his backyard in Pittsburgh. He used an RV-6 Dynascope, a 15 cm f/8 reflector. On the following night, Tom Hoffelder, formerly of the Pittsburgh group, found 103 Objects from Akron, Ohio. He was using a 25 cm f/6 reflector. On April 11–12, Tom Reiland found 103 Objects using a 15 cm f/6 reflector. In the following year, on March 10, 1978, Ed Flynn found 102 Objects with his 25 cm f/5.6 reflector.

In 1979, about fifty club members turned out for our San Jose Astronomical Association Messier Marathon on the nights of March 23/24, 24/25, 30/31 and March 31/April 1. We made it a habit to pick two consecutive weekends, on both Friday and Saturday nights, in order to make the best of the weather. Only a dozen of us actually participated in the Marathon, others stayed only until midnight before packing up and driving back home. On March 24–25 I picked up 107 Objects, missing only M74 and M110 in the evening sky and M30 in the morning sky. A few nights later, Tom Reiland of Pittsburgh also picked up 107 Objects, missing only M74, M77 and M30. On March 30/31, Gerry Rattley (Figure 3.5) and I were back on Loma Prieta, again running the Marathon. We each found 108 Objects, missing M74 and M33.

During the following year I lectured widely on the Messier Marathon, wrote articles for *Astronomy Magazine*[3] and *Night Skies* magazine,[4] recorded onto audio cassette tapes instructions on how to find each Object, and further refined the observing order and Marathon dates. I determined that from my latitude (37 degrees north), and with my horizons, one could find 109 Objects as early as March 12 (in later years I picked up 109 Objects as early as March 5 from the same location). This certainly opened up the window for the Marathon, giving us several weekends to chose from. In 1980 I located 109 Messier Objects on the night of March 12/13 without using star charts. Instead, I followed the search instructions on the cassette tapes that I had previously recorded.

Figure 3.5

Gerry Rattley at Loma Prieta, California, 1979. Picture taken by the author.

That year most clubs held the Marathon on the night of March 15/16. The members of the Amateur Astronomers of Pittsburgh set up in six inches of snow some forty miles south of Pittsburgh. Tom Reiland and Ed Flynn each found 109 Objects, missing only M30. On the same night, on the West Coast, at Loma Prieta, both Ken Wilson and I also found 109 Objects, both missing M30.

As the years progressed, the Marathon was being run from more and more locations. Astronomy clubs, especially those in the western United States, began running the Marathon as their March star party. Through the 1980s and 1990s, additional magazine articles and public lectures furthered the idea of the Marathon.

It was on March 23/24, 1985, that Gerry Rattley, observing from

Figure 3.6

Don Machholz and his 25 cm reflector, 1978. Picture taken by the author.

Dugas, Arizona, picked up M30 in the morning twilight, completing the full list of 110 Messier Objects in one night. While others have since duplicated this feat, your observing conditions may preclude such a possibility. We shall therefore turn next to an examination of the factors which determine the number of Messier Objects you can expect to see.

Factors affecting the Messier Marathon

Perhaps by now you are asking: 'How can I increase my chances of locating and identifying most, or all, of the Messier Objects in one night?' That's a good question; everyone asks it before the Marathon, and a few ask it the next morning. In the following are some of the elements affecting the Messier Marathon.

- The *date*, one of the most important factors, is determined by the day of the week, the phase of the moon, and the weather.
- Your *latitude* can limit the number of Messier Objects visible.
- Somewhat related are your *local horizon heights* and *light pollution*.
- The *instrument* you use plays a role, as does your *experience*.
- Finally, it helps to have a good *search sequence* so that you'll be looking in the right place at the right time.

Date

Every year it is the same question: 'When are we going to hold the Messier Marathon this spring?' And every year the answer is different. This is because the date of the Messier Marathon depends upon the day of the week, moon phase, and weather. We'll examine these three areas first, then look at the suggested Messier Marathon dates.

Day of the Week

Since the Marathon can last all night long, and the night generally begins at about 7:30 PM local time and ends at roughly 5:00 AM, most observers choose a weekend so that they will not have to then face a full day of work. Astronomy clubs are especially keen on this idea since they generally hold their star parties on Saturday nights anyway. There are exceptions, however. Although the Marathon is advertised as an all-nighter, it does not need to be that way. I will often find about 65 Messier Objects before 10:00 PM, then go to bed until about 2:00 AM when I arise to locate the remaining Objects. Other alternatives include choosing a Friday night/Saturday morning or taking the day off work if you typically work Monday through Friday.

Moon Phase

When the moon illuminates the Earth's atmosphere, the contrast of objects in the sky is reduced and the objects are more difficult to see. This applies more to areas near the vicinity of the moon, less so farther away. This effect is minimized when the air is thin (as at higher elevations) or when the air is clear and dry. Moreover, a bright moon ruins dark adaptation as it also illuminates the landscape, telescope, charts, and everything else surrounding the observer. Therefore, it is suggested that the Marathon be held near the time of New Moon, when the moon rises and sets with the sun.

For marathoners in the springtime, the moon is not a factor when it is between three days before New Moon and two days after. Outside of those times the moon's brightness and possible proximity to one or more of the Messier Objects is likely to create observing problems.

There are exceptions to this rule. On March 19/20, 1984, I observed 109 Messier Objects, despite the moon being less than three days past Full. I logged more than half of the Catalogue before the moon became a serious problem in the eastern sky. Then, at about 2.00 AM, I located the remaining Objects.

With the above factors in mind, I have tabulated suggested Messier Marathon dates for the next few years (see Table 3.1). The dates in Table 3.1 are only suggestions. Those not tied to weekends may choose other dates centered on the New Moon.[5] If you live in an area typically having bad weather during March, try picking several nights.

This brings us to the least predictable factor concerning the Marathon, the weather.

Weather

Because this is difficult to predict long in advance, some Marathons are scheduled for several nights. They may not even be on consecutive weekends, but rather be timed to the New Moons, roughly one month apart. This provides a good chance for at least one clear night.

Marathoners are warned that a cloudy afternoon may not result in a cloudy night. It may well be worth your time and effort to prepare for the evening, so that if it does clear up you will be ready. To begin your observing, you need only a clear sky in the west. Then, as the evening progresses, if the clouds continue to retreat you may be able to complete the list.

Related to weather is the transparency of the atmosphere. Haze, especially in the presence of moonlight or light pollution, makes the observation of objects at low altitude difficult. Messier Objects of

Table 3.1 **Proposed dates for the Messier Marathon, 2001–2050.**

These are the proposed Messier Marathon dates for the years 2001 to 2050. They are chosen with respect to the New Moon, weekends (Saturdays are selected) and the maximum number of Messier Objects visible. The New Moon dates are listed in decimals of a day, in Universal Time.

Year	New Moon date(s)	Primary weekend	Secondary weekend
2001	March 25.1	March 24	
2002	March 14.1	March 16	
2003	March 3.1, April 1.8	March 29	March 08
2004	March 21.0	March 20	
2005	March 10.4	March 12	
2006	March 29.5	March 25	April 01
2007	March 19.1	March 17	March 24
2008	March 7.8, April 6.2	March 08	April 05
2009	March 26.7	March 28	March 21
2010	March 15.8	March 13	March 20
2011	March 4.8, April 3.6	March 05	April 02
2012	March 22.6	March 24	March 17
2013	March 11.8	March 09	March 16
2014	March 1.3, March 30.8	March 29	March 01
2015	March 20.4	March 21	
2016	March 9.1, April 7.5	March 06	April 02
2017	March 28.1	March 25	
2018	March 17.5	March 17	
2019	March 6.7, April 5.4	March 09	April 06
2020	March 24.4	March 21	March 28
2021	March 13.4	March 13	
2022	March 2.7, April 1.2	March 05	April 02
2023	March 21.7	March 18	March 25
2024	March 10.4	March 09	
2025	March 29.4	March 29	
2026	March 19.1	March 21	March 14
2027	March 8.4, April 7.0	March 06	April 03
2028	March 26.2	March 25	
2029	March 15.2	March 17	March 10
2030	April 2.9	March 30	
2031	March 23.1	March 22	
2032	March 11.7	March 13	
2033	March 1.2, March 30.7	March 05	April 02
2034	March 20.4	March 18	
2035	March 10.0	March 10	
2036	March 27.8	March 29	
2037	March 17.0	March 14	March 21

Table 3.1 (cont.)

Year	New Moon date(s)	Primary weekend	Secondary weekend
2038	March 5.5, April 4.7	March 06	April 03
2039	March 24.6	March 26	March 19
2040	March 13.1	March 10	March 17
2041	March 2.6, April 1.0	March 30	March 02
2042	March 21.7	March 22	
2043	March 11.3	March 14	
2044	March 29.4	March 26	
2045	March 18.7	March 18	
2046	March 7.8	March 10	March 03
2047	March 26.5	March 23	March 30
2048	March 14.6	March 14	
2049	March 4.0, April 2.5	March 06	April 03
2050	March 23.5	March 19	March 26

low surface brightness such as M74, M33 and M110 can fade to invisibility.

Latitude

I have learned through experience that your latitude makes a big difference in what you can and cannot see during the Messier Marathon. All other things being equal, an observer at +25 degrees latitude will see at least as many Messier Objects as an observer 15 degrees farther north, and, in most cases, more. One will also have more nights to view 108, 109 or 110 Objects than will a northern counterpart. There is no way around this; your latitude determines the maximum number of Messier Objects that you can see in one night.

Observers between latitudes +35 and +50 degrees may favor mid-March to late March, when the maximum number of Messier Objects can be seen. Prior to mid-March, the Messier Objects in the evening are easy to see. But some of those rising in the east never get quite high enough to see before twilight brings the night's observing to an end. With each passing day, however, they are slightly higher before twilight, and therefore easier to see. As we approach April, only the globular cluster M30 is difficult to see in the morning twilight. Meanwhile, in the evening sky, things are getting more difficult each night as galaxies M74, M77, M33 and M110 disappear in the evening twilight.

Observers located between +20 and +35 degrees latitude will be particularly interested in dates during late March, when all 110 Messier Objects should be visible. Their window for 109 Objects is large too, beginning in early March and extending through the month.

As one moves further south, the morning objects become visible sooner while the evening objects disappear sooner. The best Marathon nights arrive earlier in March. As you move south of the equator, the northern Messier Objects disappear in the evening and morning twilights, and the optimal time for the Messier Marathon shifts from March to mid-May.

The hide and seek of various Messier Objects from different latitudes can be best demonstrated by the following horizon charts, Charts 3.2–3.22, designed and drawn by the Deep Space 3D™ program.[6] They are shown for various latitudes and dates. For each, the curved line shows the western horizon at sunset on the far right of the chart. The line just to the left of this is the horizon at astronomical twilight; this is very important because objects near this line are setting as the sky darkens. They will be visible, but with some difficulty. On the far left we have the eastern horizon lines for morning astronomical twilight and (to the left of that) sunrise. Objects to the left of the astronomical twilight line will be nearly impossible to see. On most charts you will find M30 there.

Just because an object is between the astronomical twilight lines does not mean it will necessarily be visible to the marathoner. Faint diffuse objects such as M74, M33 and M110 disappear a few days before 'their time', while the condensed M77 may be visible longer. In the morning sky a bright M30 may be visible a few days before it crosses the 'morning twilight' line into dark sky.

Local horizons

High local horizons, especially to the west–northwest and southeast, can block off some of the difficult objects that are near the horizon around twilight. Sometimes you can trim a tree or pick up your telescope and move it a short distance to a better location. Other horizon problems are not as easily solved. If your western horizon is high, then choose an earlier date so you can locate the

+55 degrees latitude

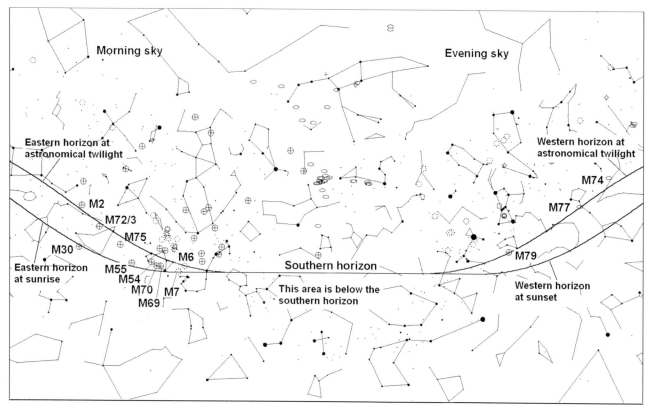

Chart 3.2

The all-night sky as seen from 55° north latitude on March 24. In the evening sky, M74, M77 and M79 are difficult to see. In the morning sky, M30 rises at sunrise, while M75, M55, M54, M69, M70 and M7 rise after morning astronomical twilight but before sunrise. They will be very difficult or impossible to see. M2, M72 and M73 rise near morning astronomical twilight, making them difficult too.

evening objects before they set. Your morning hunting will suffer, though. Likewise, if your eastern horizon is high, pick a later date so that the objects will have more time to rise above the local horizon before morning twilight.

Light pollution

Much as moonlight washes out the sky, so too does excessive man-made light. This is especially true if the light source is in the east or west, where finding the fringe Messier Objects is most difficult. If it is a neighbor's light, you may politely ask them to turn it off. At one public park I made covers out of large plastic bags to shield the lights. Other solutions include cupping your hand over your non-observing eye, or using an eye patch, or placing a small blanket or towel over your head and eyepiece.

Zodiacal light can also be a problem, especially while finding M74 in the evening or M72 and M73 in the morning. You can't do much about that, except perhaps to increase magnification to darken the background.

41

+50 degrees latitude

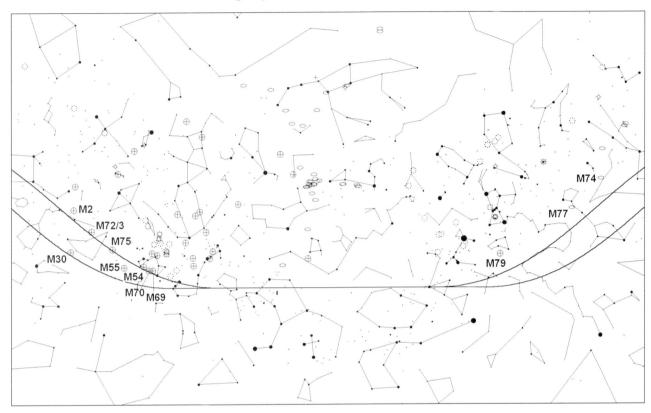

Chart 3.3

The all-night sky as seen from 50° north latitude on March 18. The evening Objects are easy to see, but in the morning sky M30 rises just before the sun, while M55 rises between astronomical twilight and sunrise. Rising near morning twilight are M2, M72, M73, M75, M54, M69, M70 and M7.

+50 degrees latitude

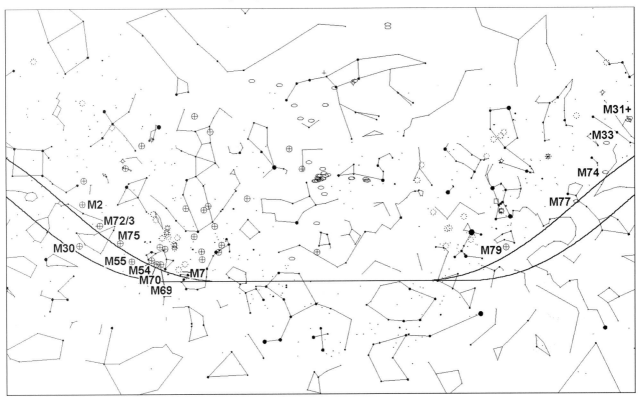

Chart 3.4

The all-night sky as seen from 50° north latitude on March 27. In the evening sky M77 is difficult to see, as are M74 and M79. In the morning sky M30 rises just before sunrise, as does M55. Other morning Objects are easier to find than they were nine days before (Chart 3.3).

+45 degrees latitude

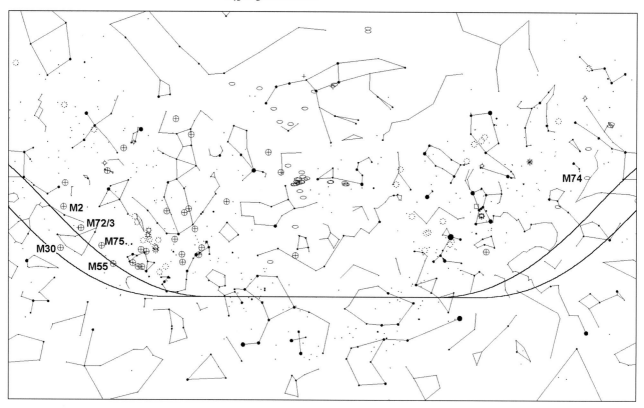

Chart 3.5

The all-night sky as seen from 45° north latitude on March 14. The only problem here is finding those morning sky Objects. M30 is impossible, while M55 rises at astronomical twilight. Also difficult are M2, M72, M73 and M75.

+45 degrees latitude

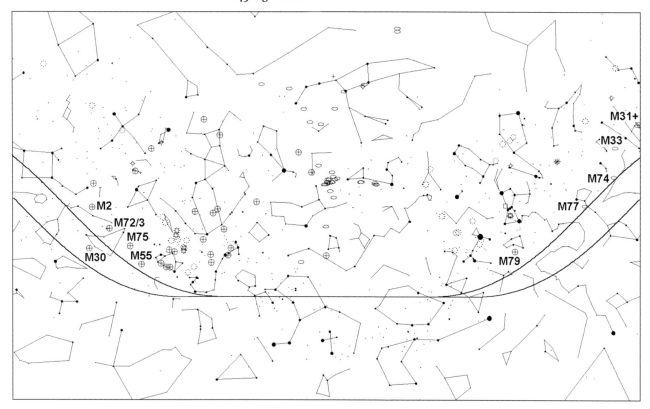

Chart 3.6

The all-night sky as seen from 45° north latitude on March 29. M74 and M77 are becoming lost in the evening sky, while only M30 is impossible in the morning sky. Over the next few nights the evening Objects become more difficult to pick up.

+ 40 degrees latitude

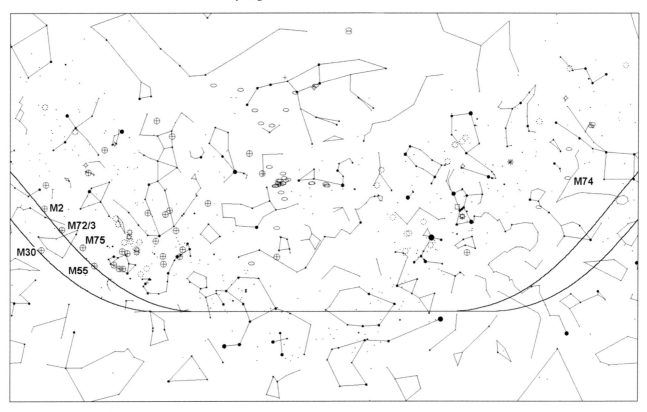

Chart 3.7

The all-night sky as seen from 40° north latitude on March 1. The only problem here is finding those morning sky Objects. M30 is impossible, while M55 rises at astronomical twilight. Also difficult are M2, M72, M73 and M75.

+40 degrees latitude

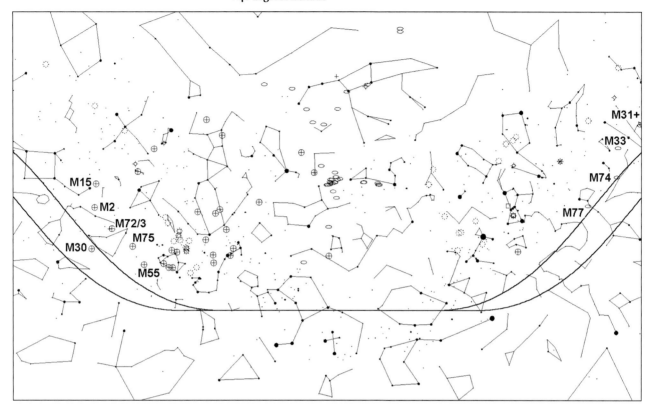

Chart 3.8

The all-night sky as seen from 40° north latitude on March 30. M74 and M77 are becoming lost in the evening sky, while only M30 is impossible in the morning sky. Over the next few nights the evening Objects become more difficult to pick up.

+35 degrees latitude

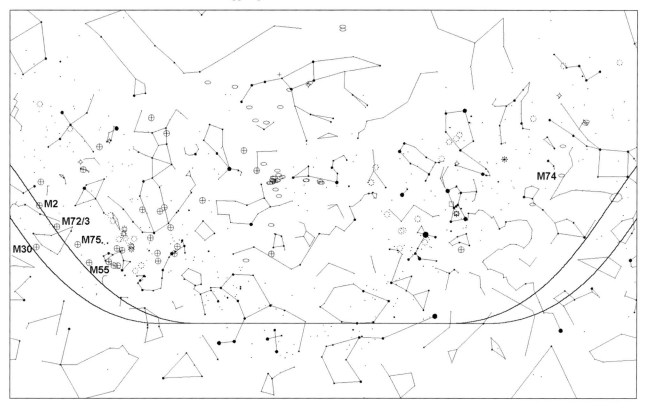

Chart 3.9

The all-night sky as seen from 35° north latitude on February 26. Only M30 is impossible in the morning sky; meanwhile M2, M72, M73 and M55 are challenging. The evening sky is easy. On days prior to this, the morning Objects are more difficult to see.

+35 degrees latitude

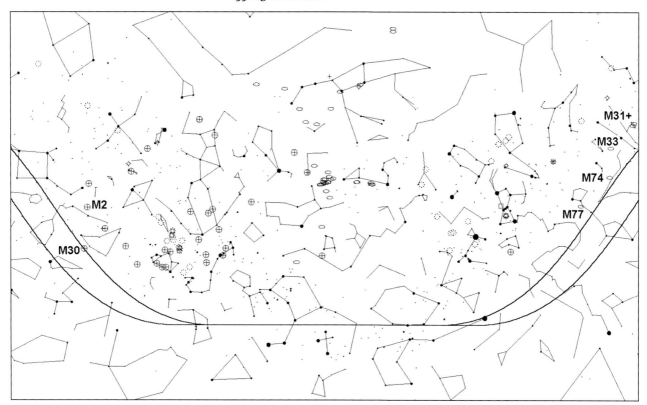

Chart 3.10

The all-night sky as seen from 35° north latitude on March 27. By now M30 is just visible in the morning sky, but soon M74 and M77 will become lost in the evening sky.

+30 degrees latitude

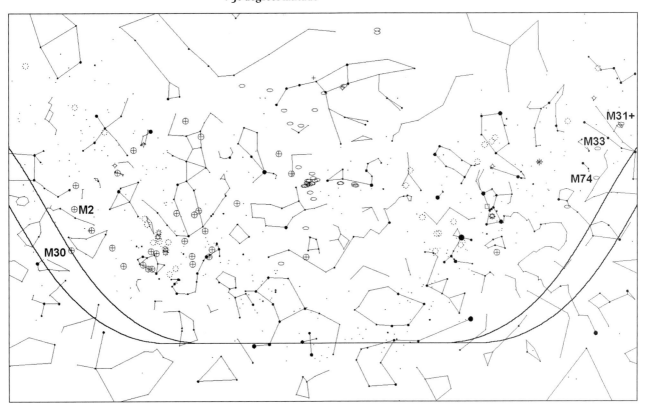

Chart 3.11

The all-night sky as seen from 30° north latitude on March 20. We now have all 110 Objects visible as M30 finally rises at astronomical twilight.

+ 30 degrees latitude

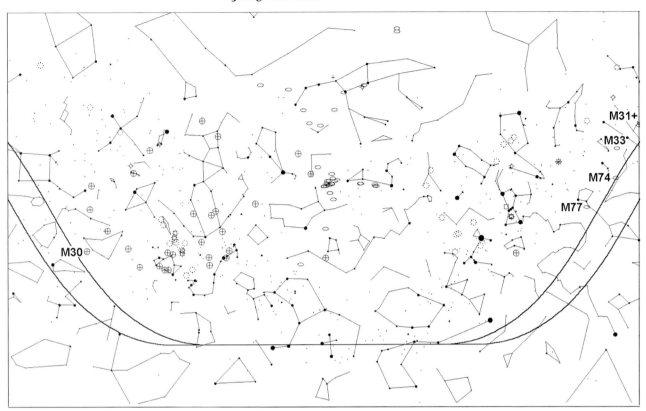

Chart 3.12

The all-night sky as seen from 30° north latitude on March 30. Although M30 is becoming easier to see in the morning sky, M74, M33 and M77 are beginning to fade in the evening sky.

+ 25 degrees latitude

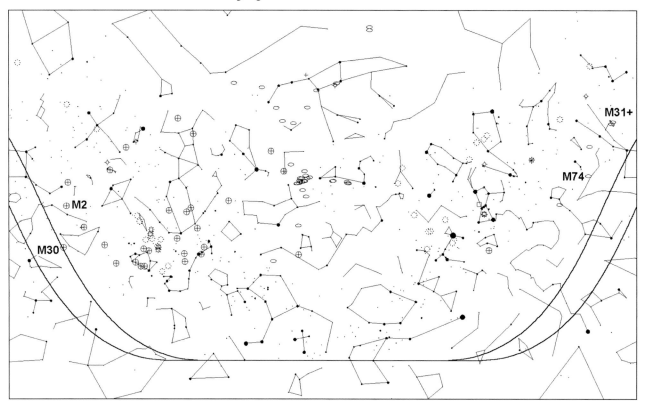

Chart 3.13

The all-night sky as seen from 25° north latitude on March 15. We now have all 110 Objects visible as M30 finally rises at astronomical twilight.

+25 degrees latitude

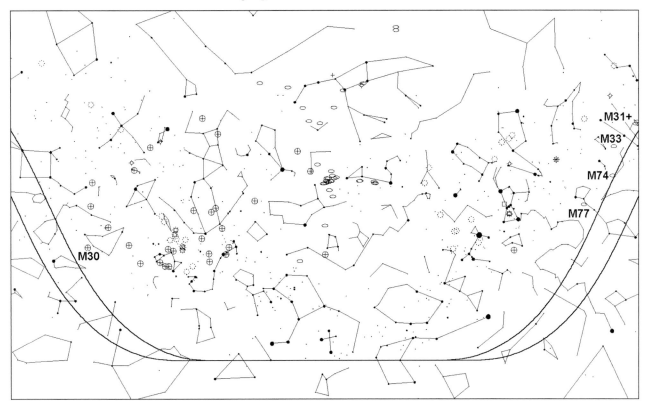

Chart 3.14

The all-night sky as seen from 25° north latitude on March 29. Although M30 is becoming easier to see in the morning sky, M74, M33 and M77 are beginning to fade in the evening sky.

+ 20 degrees latitude

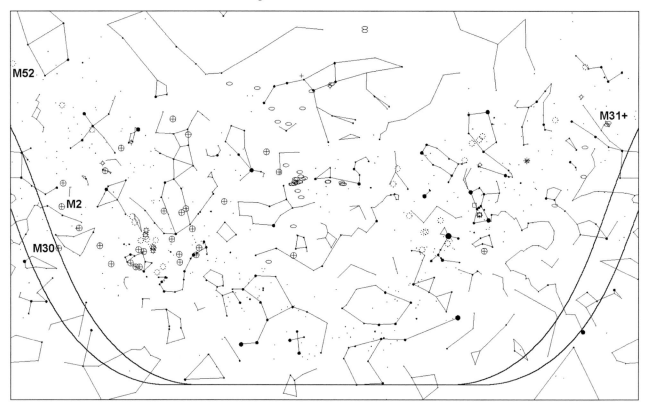

Chart 3.15

The all-night sky as seen from 20° north latitude on March 11. Before mid-March you can see all 110 Messier Objects in one night.

+ 20 degrees latitude

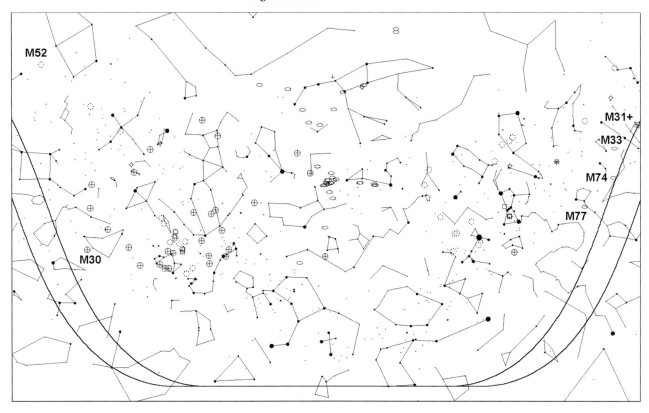

Chart 3.16

The all-night sky as seen from 20° north latitude on March 28. Near the end of March we lose M31, M32, M110 and M74 in the evening twilight.

+10 degrees latitude

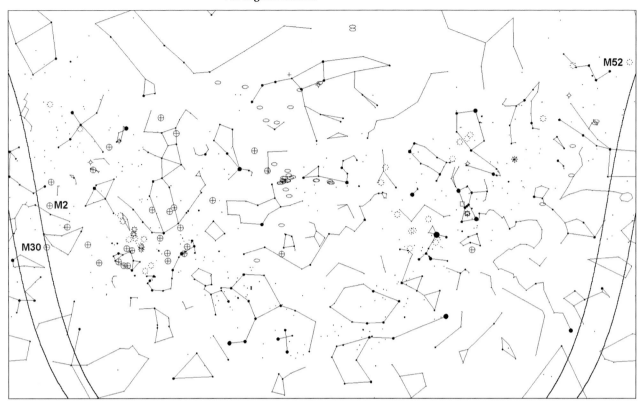

Chart 3.17

The all-night sky as seen from 10° north latitude on March 5. In early March all 110 Messier Objects are visible. In the evening sky M52 is difficult to see.

+10 degrees latitude

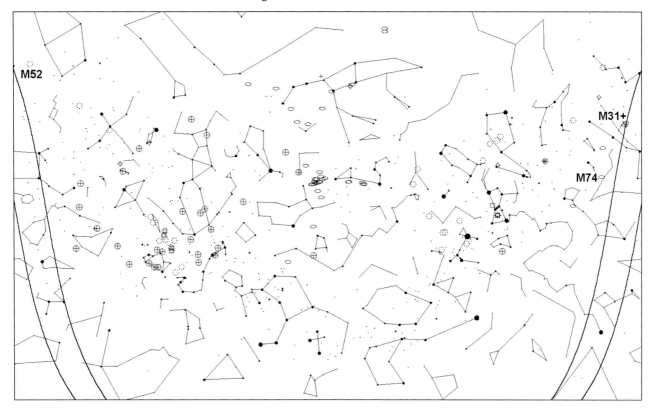

Chart 3.18

The all-night sky as seen from 10° north latitude on March 20. From this southerly location, M31, M32 and M110 are now the first to disappear in the evening sky.

From the equator

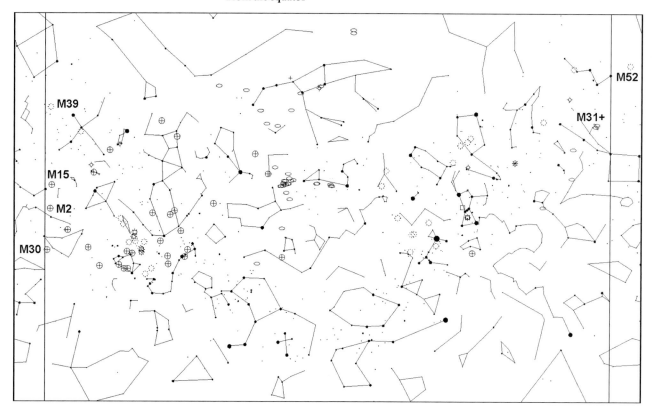

Chart 3.19

The all-night sky as seen from the equator on **March 2**. Only M52 is invisible from this southerly location.

From the equator

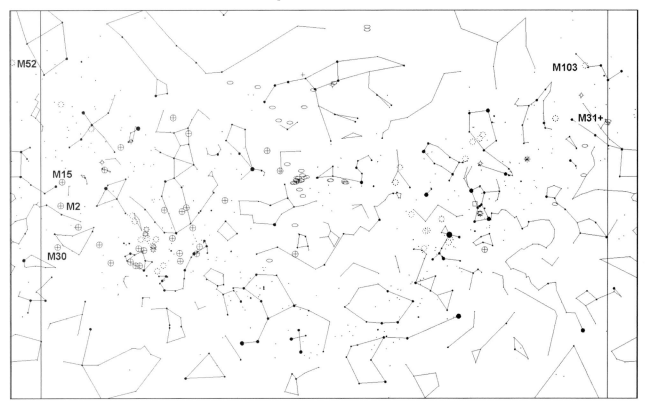

Chart 3.20

The all-night sky as seen from the equator on March 11. M52 is still hidden, but now M31, M32 and M110 begin to disappear in the evening sky.

−10 degrees latitude

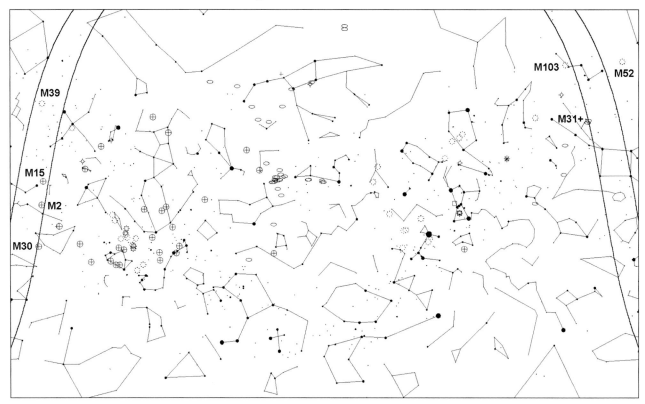

Chart 3.21

The all-night sky as seen from 10° south latitude on February 27. For a few days in late February, up to 106 Messier Objects can be seen before M31, M32, and M110 sink in the west.

−10 degrees latitude

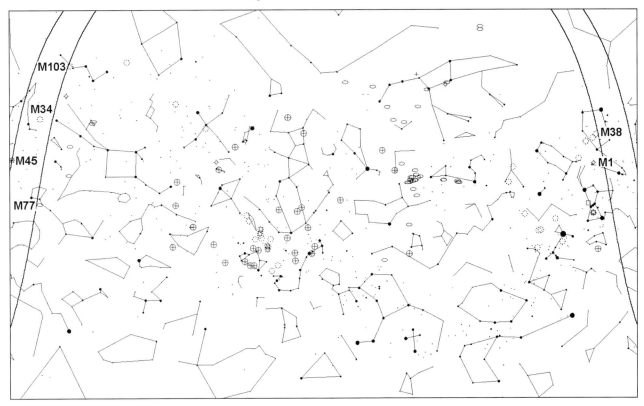

Chart 3.22

The all-night sky as seen from 10° south latitude on May 20. For a few days in late May, 108 Messier Objects can be seen; only M45 and M34 are missing.

Table 3.2 **The number of Messier Objects visible from various latitudes from mid-February to mid-April.**

Date	55N	50N	45N	40N	35N	30N	25N	20N	10N	EQU	10S	Date
2/16	97	100	102	104	104	106	105	105	105	104	102	2/16
2/17	97	100	102	104	105	106	106	105	105	105	102	2/17
2/18	97	100	102	104	105	106	106	105	106	105	103	2/18
2/19	98	100	102	104	105	106	106	107	107	105	104	2/19
2/20	99	100	103	104	106	106	107	107	107	105	105	2/20
2/21	99	100	104	104	106	107	107	108	107	105	105	2/21
2/22	99	100	104	105	106	109	108	108	107	105	105	2/22
2/23	99	100	104	105	106	109	108	108	108	105	105	2/23
2/24	99	100	104	106	108	109	108	108	108	105	105	2/24
2/25	99	100	105	106	108	109	109	109	108	105	105	2/25
2/26	99	100	105	106	109	109	109	109	109	105	105	2/26
2/27	99	101	105	107	109	109	109	109	109	107	106	2/27
2/28	99	101	105	107	109	109	109	109	109	109	105	2/28
3/1	99	101	105	109	109	109	109	109	109	109	104	3/1
3/2	100	101	106	109	109	109	109	109	109	109	104	3/2
3/3	100	102	106	109	109	109	109	109	109	109	105	3/3
3/4	100	102	107	109	109	109	109	109	109	109	105	3/4
3/5	100	102	107	109	109	109	109	109	110	109	104	3/5
3/6	100	102	107	109	109	109	109	109	110	109	104	3/6
3/7	100	102	108	109	109	109	109	109	110	109	104	3/7
3/8	100	103	108	109	109	109	109	109	110	109	104	3/8
3/9	100	103	108	109	109	109	109	109	110	109	104	3/9
3/10	100	103	108	109	109	109	109	109	110	109	104	3/10
3/11	100	103	108	109	109	109	109	110	110	109	104	3/11
3/12	100	104	108	109	109	109	109	110	110	106	105	3/12
3/13	100	105	108	109	109	109	109	110	110	106	105	3/13
3/14	100	105	108	109	109	109	109	110	110	106	104	3/14
3/15	100	106	109	109	109	109	110	110	110	106	104	3/15
3/16	100	107	109	109	109	109	110	110	110	106	104	3/16
3/17	100	108	109	109	109	109	110	110	110	106	104	3/17
3/18	100	108	109	109	109	109	110	110	110	106	104	3/18
3/19	100	108	109	109	109	109	110	110	110	106	104	3/19
3/20	100	108	109	109	109	110	110	110	110	106	103	3/20
3/21	100	108	109	109	109	110	110	110	110	106	103	3/21
3/22	100	108	109	109	109	110	110	110	109	106	103	3/22
3/23	100	108	109	109	109	110	110	110	107	106	103	3/23
3/24	100	108	109	109	109	110	110	110	107	106	102	3/24
3/25	98	108	109	109	109	110	110	110	107	106	102	3/25
3/26	98	108	109	109	109	110	110	110	107	105	102	3/26

Table 3.2 (cont.)

Date	55N	50N	45N	40N	35N	30N	25N	20N	10N	EQU	10S	Date
3/27	99	108	109	109	110	110	110	110	107	103	102	3/27
3/28	99	108	109	109	110	110	110	110	107	103	102	3/28
3/29	99	107	109	109	110	110	110	110	106	102	102	3/29
3/30	98	107	109	109	110	110	110	107	106	103	102	3/30
3/31	98	106	107	108	109	109	109	106	106	103	102	3/31
4/1	98	106	107	108	109	109	109	106	105	103	102	4/1
4/2	98	106	107	108	109	109	109	107	105	103	102	4/2
4/3	98	105	107	107	109	109	109	107	105	103	102	4/3
4/4	98	105	107	107	109	109	109	109	105	103	102	4/4
4/5	98	105	107	107	108	109	109	108	105	103	101	4/5
4/6	98	105	107	107	108	108	109	108	105	103	101	4/6
4/7	98	105	107	108	108	108	108	108	105	103	101	4/7
4/8	98	105	107	108	108	108	107	108	105	103	101	4/8
4/9	98	105	107	108	108	107	107	108	105	103	101	4/9
4/10	98	105	106	108	108	107	107	107	104	103	101	4/10
4/11	97	105	106	108	107	107	107	107	104	103	101	4/11
4/12	97	105	106	107	107	107	107	107	104	103	101	4/12
4/13	97	105	106	107	107	107	107	107	105	103	101	4/13
4/14	97	105	105	107	107	107	107	107	105	101	101	4/14
4/15	95	105	105	107	107	107	107	107	105	101	101	4/15
4/16	94	105	105	106	107	107	107	107	105	101	100	4/16
Date	55N	50N	45N	40N	35N	30N	25N	20N	10N	EQU	10S	Date

Instrument

I've always said that what you can see will depend upon three things: your eyes, the sky and your telescope, in that order. We've discussed the sky, and in the next section we'll look at your eyes. Here we'll discuss the instrument.

Almost any telescope today is better than the instruments used by Charles Messier when he compiled his Catalogue. But Messier did not have to search for M74 when it was 10 degrees above the horizon or M72 and M73 in morning twilight. Generally, you can see better with a larger aperture (because the image is larger, all other factors being equal) while a longer focal ratio produces better contrast. The tradeoff is that such instruments have smaller fields, making it harder to find objects if sweeping is involved.

Popular among Marathoners are 20 cm (objective size) Schmidt–Cassegrains, refractors in the 50 mm to 150 mm size, and reflectors in the range of 10 to 50 cm. At recent All-Arizona Messier Marathons, telescopes with apertures 25 cm or greater made up half the number of telescopes. The largest category, however, was those with a 20 cm aperture, comprising 33% of the eighty-eight instruments.[7] Astronomers usually use what they have, which is advisable since the observer is most familiar with his or her own telescope. No matter what the instrument, make sure that the optics are clean and collimated.

The mounting also makes a difference. If your equatorial mount is aligned to the North Pole, you need only the written instructions in Part 2 of this book to marathon. Offsets are provided to get you from the previous object or a bright star to the next object. If you have a Dobsonian mount, don't set up to the east or west of a tall telescope: you won't be able to see over it. If your telescope is on an altazimuth mount, you may have trouble finding objects overhead.

An increasing number of astronomers use binoculars for the Messier Marathon. Some use 7×35 or 7×50 binoculars to sharpen their skills by attempting to find all the objects that would be visible in a telescope. Others use binoculars because they are easier to use than most telescopes and can be directed around the sky more quickly than larger instruments.

Let me say a few words about electronic setting circles. Referred to as GOTO telescopes because they 'Go To' wherever, you want, this equipment is attached to your telescope mount which automatically slews your telescope to any location or object you desire. Can they be used for the Messier Marathon? Sure they can, but the emphasis of the Marathon now shifts to seeing the Messier Objects, rather than both locating and seeing them. One should not expect those finding the Messier Objects using the old-fashioned methods of star offset or star hopping to compete against these electronic wonders. A search sequence for electronic setting circles is presented in Table 3.3.

At present, there is one telescope manufacturer (Meade) which has written a Messier Marathon program that can be downloaded from their web site to your GOTO computer. It automatically slews your telescope from one Messier Object to another, pausing at each one. The current search sequence lists the objects by right ascension, not necessarily the best sequence for the Marathon.

Table 3.3 **Search sequence for GOTO telescopes.**

#	Object	#	Object	#	Object	#	Object
1	M77	29	M95	57	M90	85	M62
2	M74	30	M96	58	M88	86	M6
3	M33	31	M105	59	M91	87	M7
4	M31	32	M65	60	M58	88	M11
5	M32	33	M66	61	M59	89	M26
6	M110	34	M81	62	M60	90	M16
7	M52	35	M82	63	M49	91	M17
8	M103	36	M97	64	M61	92	M18
9	M76	37	M108	65	M104	93	M24
10	M34	38	M109	66	M68	94	M25
11	M45	39	M40	67	M83	95	M23
12	M79	40	M106	68	M5	96	M21
13	M42	41	M94	69	M13	97	M20
14	M43	42	M63	70	M92	98	M8
15	M78	43	M51	71	M57	99	M28
16	M41	44	M101	72	M56	100	M22
17	M93	45	M102^a	73	M29	101	M69
18	M47	46	M53	74	M39	102	M70
19	M46	47	M64	75	M27	103	M54
20	M50	48	M3	76	M71	104	M55
21	M48	49	M98	77	M12	105	M75
22	M1	50	M99	78	M10	106	M15
23	M35	51	M100	79	M14	107	M2
24	M37	52	M85	80	M107	108	M72
25	M36	53	M84	81	M9	109	M73
26	M38	54	M86	82	M4	110	M30
27	M44	55	M87	83	M80		
28	M67	56	M89	84	M19		

Note:
^{a} M102 = NGC 5866.

Experience

If you can, right now, tell the person nearest you what M1 looks like through your telescope, or what type of object M107 is, or which Messier Object is close to M65, then you have more than enough experience to do well in the Messier Marathon. If you can't, then the best way to gain the experience is to go out tonight and see some Messier Objects. This is not the type of thing you can learn from a book (even this one), but comes from nights out under the stars. As you learn more, your eyes will also learn to see more. Practice using averted vision (concentrating on the object, but looking to the side of it). Log a few splendors each night. Before too long you will become a 'seasoned' observer.

Search sequence

A well-designed search sequence should take you from one Messier Object to the next with minimum movement; it should also get you through the list without missing anything. This means it is important to find evening Objects before they set, and morning Objects just after they rise. One of the best ways to prepare for the Marathon is to sit down with a star map, planisphere and/or astronomy computer program, and develop your own search sequence

The search sequence used in Part 2 of this book will do this for you, but it is by no means the only possible sequence. While it will work for most mid-northern latitudes with low horizons during March, your site may not be ideal. A tall tree or light pollution in the northwestern sky may cause you to move M52 and M103 from the evening list to your morning list. Holding the Marathon in late March or early April may force you to move M31, M32 and M110 onto the morning list too.

Messier Marathons during other times of the year

Even though March and April are the best months for logging the greatest number of Messier Objects, a Marathon can be held during other times of the year too. For these Marathons the sun is positioned among a different set of Messier Objects and the challenges at evening and morning twilights are of a different nature than those we find in the springtime.

For any latitude, the number of visible Messier Objects can be calculated using a computer program written by Bev M. Ewen-Smith. It can be downloaded free from the web site: www.ip.pt/coaa/messmara.zip. Please note that this program treats M102 as if it were M101, so it will tell you that the maximum number of Objects is 109.

We have become familiar with the traditional March Messier Marathon. Charts 3.23–3.27 show the sky from various latitudes in the October/November timeframe. That is when the sun is near

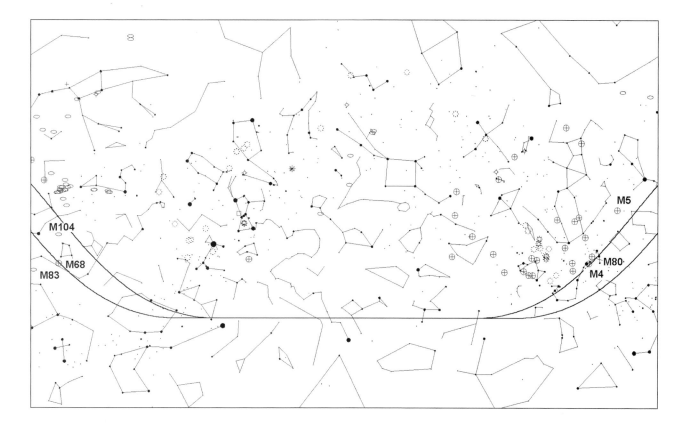

Chart 3.23

The all-night sky as seen from 40° north latitude on October 20. For two weeks, 107 Messier Objects can be seen, with very busy searches at both evening and morning twilight.

14 hours of right ascension and it blocks out very few of the Messier Objects. It is also south of the equator, giving longer nights to northern hemisphere observers. In March the globular cluster M30 is difficult to see. In the fall Messier Marathon M30 is easy, and the southern galaxy M83 replaces M30 as the hardest to find.

The search sequence changes too. The summer Milky Way Objects are bunched in the evening sky; they must be seen as twilight falls. The marathoner is much busier at evening twilight at this time of year than in March. The pace slows down as one continues through the Objects higher in the sky. The winter Milky Way is also sparsely populated, with few Messier Objects appearing in the eastern sky. Near midnight there is little new to see. A couple of hours before morning twilight the galaxies are rising. After the northern ones are seen, the observer turns to the mid-northern declinations to finish up the Virgo group. It is critical that your southwest and eastern horizons are low and dark.

Finally, I include the sky from the equator on May 23 (Chart 3.28). Here is a short window where 109 Objects can be seen. This is nearly three months after the early March window also yielded 109 Objects.

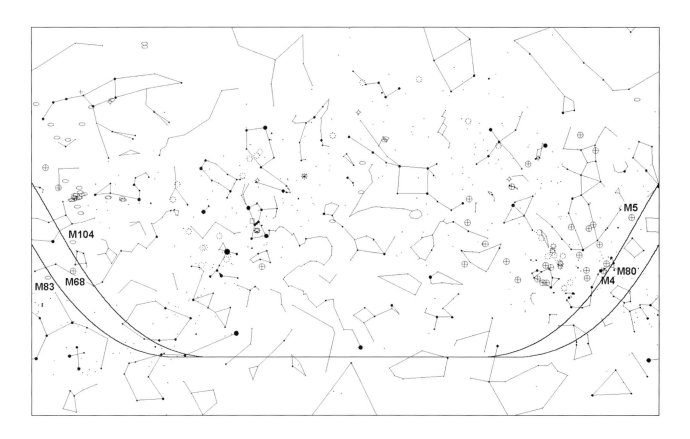

Chart 3.24

The all-night sky as seen from 30° north latitude on October 28. With good horizons, as many as 108 Messier Objects should be visible.

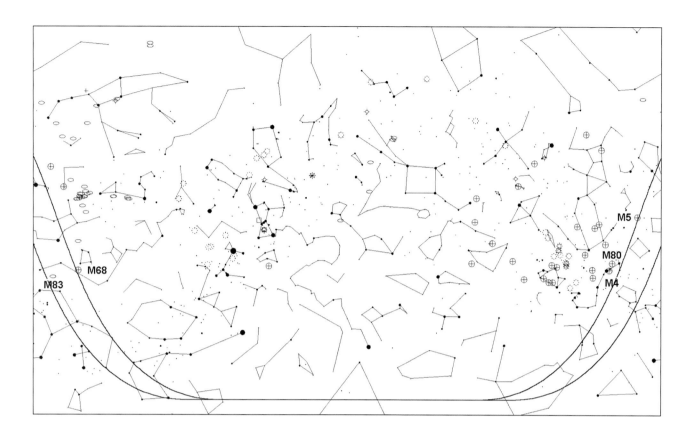

Chart 3.25

The all-night sky as seen from 20° north latitude on October 30. During several days, 109 Messier Objects can be seen, with only M83 not visible.

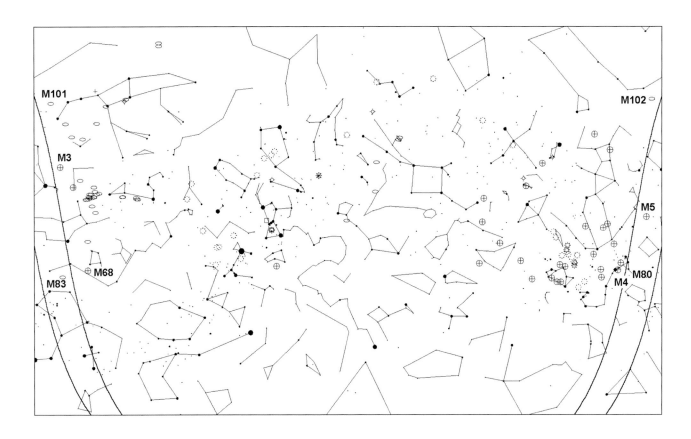

Chart 3.26

The all-night sky as seen from 10° north latitude on November 4. For about one week, centered on this date, 108 Messier Objects should be visible.

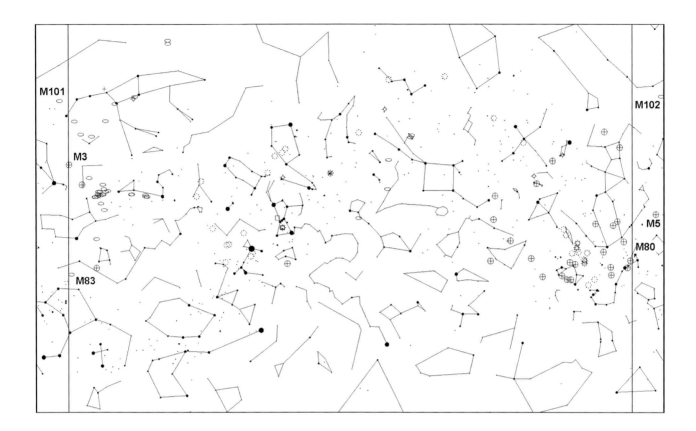

Chart 3.27

**The all-night sky as seen from the equator
on November 8**. As we move south, we lose
both M101 and M102, plus M5, reducing
our possible total to 107 visible Objects.

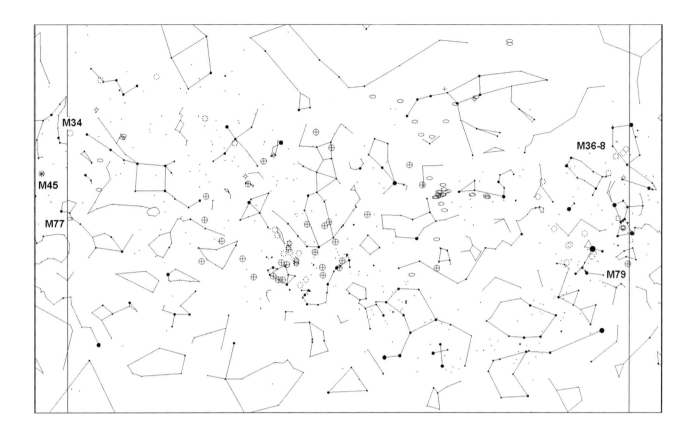

Chart 3.28

The all-night sky as seen from the equator on May 23. In late May there are a few days when 109 Messier Objects (all except M45) are visible.

Getting your astronomy club involved

Many observers are introduced to the Messier Marathon through their astronomy club. The club often conducts the Marathon as the first star party of the season. These gatherings bring the members together for a night of observing with a mutual goal. Some of the astronomy associations have become quite involved: observers register beforehand and pay a small fee to sponsor awards. These awards are presented to those finding the most Objects or those who beat their personal best, or, say locate and observe more than one hundred Objects. Rules are simple and few: confirmation is left to the observer (honor system), and those using electronic setting circles (GOTO systems) are placed in a different category. Some groups ask for a short description of each Object along with the time that it was located. Your club need not be as comprehensive as this, but, for further reference, at least try to record the name of each observer, instrument and the number of Messier Objects observed.

How do you get started? Pick a date for the Marathon using Table 3.1. Some clubs pick two or four dates, perhaps a Friday and

Figure 3.7

An astronomy club star party is an ideal
opportunity to conduct a Messier Marathon.

Figure 3.8

Anticipation grows as the sun sets in the western sky.
Both images courtesy of Steve Coe.

Saturday night of one weekend, or two consecutive weekends. Another alternative is to pick two weekends on the New Moons, about a month apart. Next, clear the dates with your executive board, and place it on the club calendar. Write an article about it for the club newsletter, and talk it up. Somewhere along the way you may find some opposition to the Messier Marathon. This is not new, and for some time controversy concerning the Marathon ran through the pages of *Deep Sky Magazine*.[8]

The main complaint is that rushing through the Messier List does not allow time to study each object. Such criticism can be ignored, since the Messier Marathon is not designed for everyone. The critic can spend the night looking at a shorter list of wonders. A counterpoint to this resistance is that the marathoner will see nearly all the Messier Catalogue in one night – many amateur astronomers never see the whole Catalogue in their entire lifetimes.

Additionally, one's searching and locating skills, necessary in most aspects of astronomy, are sharpened during the Marathon. The benefit of seeing, in one night, the major building blocks of our galaxy: open and globular clusters, diffuse and planetary nebulae, along with other galaxies, cannot be ignored. Finally, there is the satisfaction of working with others toward a common goal, and then finally achieving it.

Choose a Marathon site that is accessible, dark, has low horizons, and is large enough to accommodate everyone. It is not easy to find such a place, but a good way to begin is to look at the club's usual star party sites, parks operated by county, state or regional districts, and private land (obtain permission). Airports and county dumps may be options in your location. Depending upon the size of the lot, you may have to limit the choicest areas of the site to the marathoners.

Realize, as I did with some disappointment in our early days, that many who come to the star party will choose not to run the Marathon. Some amateur astronomers come out just to watch the marathoners! Many will pack up and go home around midnight. That is OK. They can still have an enjoyable time. In advance, I would type up a list of additional objects visible throughout the night: planet, comet and asteroid positions, along with moon data and times of twilight. This information is helpful to everyone and helps to tie the group together. Finally, keep a positive attitude. People are here to learn and have fun; they will be back next year if they think they are getting something out of it. And don't forget to write a short summary article for your club newsletter.

When will the next Marathon be? The next 'regular' Messier Marathon will be held next spring. But for those who have advanced beyond this, there are always other types of marathons involving other lists of objects. And for those not yet ready for the full Messier Marathon, try a mini-Messier Marathon four times a year to help improve the observing skills of all the club members. You will find these topics discussed in the next chapter.

Hints for running the Marathon

Running the Messier Marathon can be an enjoyable experience, one to be remembered for many years. Some observers conduct the Marathon each and every year, attempting it (weather permitting) on different dates and often with various instruments and search sequences.

Here are some suggestions to help you make the most of Marathon Night. The key to conducting a worthwhile Marathon is to know your scope and the sky. It is helpful to practice before the Marathon date, observing a few objects each night. Experience has shown that the following objects are the most difficult to find in the evening: M74, M33, M110 and M77. In the morning M30 is missed most, mainly because it usually rises during twilight. Also difficult in the morning are M55, M72, M73 and M2.

Pick the right date and know what to expect. Avoid the moon and be aware of optional dates if foul weather occurs. Make sure that your instrument has clean optics and is collimated. The axes on the mount should move freely. Bring eyepieces yielding both low power for finding the Objects, and high power for seeing the smaller and more difficult Objects. Know the field of view of both your finder and main scope at low power. Align the finder scope to the main scope during daylight or twilight.

Be sure to bring warm clothes, food, extra flashlight batteries and pencils. Include lists and maps of non-Messier Objects (and the planets) to observe. Construct a logsheet, such as that shown on p. 78. You may wish to try for the Omega Centauri Globular Cluster, which culminates around 2:00 AM in late March. Arrive in plenty of time. Setting up your instrument and becoming adapted to twilight will probably take longer than you expect. You can begin by finding

Figure 3.9

The northwestern sky at evening twilight, as seen from mid-northern latitudes. Photographed and labeled by the author.

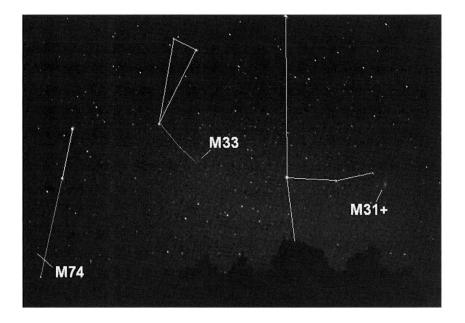

your first Object (M77) about an hour after sunset, ten or fifteen minutes before astronomical twilight.

A common problem arising during the Marathon is dew or fog forming on the eyepiece. (A long dew cap should prevent this problem from occurring on the objective.) This is caused when the eyepiece, chilled by the air, comes into contact with the warm air emitted by you when you look through the eyepiece. One solution is to electrically warm the eyepiece. Another is to keep the eyepiece end of the telescope covered when not in use. Some observers use two nearly identical eyepieces: keeping one warm while using the other. When it fogs up, they exchange it. Too many Marathoners fail in the final half-hour because they run out of time – twilight washes out the sky before they can find all the objects in the east. Don't let this happen to you.

Know when twilight occurs. Allow time to locate each object. Don't spend so much time on one Object that you miss out on finding others. In the evening M74 and M33 can cause this to happen. This is why they are sequenced after the faster-setting and more easily visible M77. Figure 3.9 shows how these Objects are configured in the sky.

Finally, have fun. Enjoy the search. If you don't reach your goals, you still have gained the experience of finding what you did find. It will be easier next time. And 'next time' may be tomorrow, next weekend, or next year. I know one observer who intentionally didn't find the final four Objects so that he would have 'something to aim for next year'.

Notes

1 Houston, Walter Scott, 'Deep-Sky Wonders', *Sky and Telescope*, March 1979, pp. 315–317.
2 Houston, Walter Scott, personal letter to Don Machholz, April 25, 1979.
3 Machholz, Don, 'Notes on a Messier Marathon', *Astronomy Magazine*, March 1980, pp. 26–28.
4 Machholz, Don, 'Notes on a Messier Marathon', *Night Skies*, December 1979, pp. 169–175.
5 Sidell, Bruce, *Moonrise Program*, Version 3.5. e-mail: bsidell@iserv.net; web site: http://www.iserv.net/mooonrise.htm.
6 Chandler, Dave, Deep Space 3D™ computer program, by Dave Chandler Co., P.O. Box 999, Springville, CA. 93265. Shareware available at http://www.DavidChandler.com.
7 Crayon, A.C., personal communication, June 20, 2001.
8 Levy, David, 'Messing Around in the Messier Catalogue', *Deep Sky Magazine* #23, Summer 1988, pp 36–37. The Marathon comments were apparently added by an editor of the magazine.
   Storch, Sam, 'Letter To The Editor', *Deep Sky Magazine* #24, Autumn 1988, p. 8.
   Lorenzin, Tom, 'Letter To The Editor', *Deep Sky Magazine* #24, Autumn 1988, p. 8.
   De Vries, G. Jan, 'Letter To The Editor', *Deep Sky Magazine* #28, Autumn 1989, p. 9.
   Suiter, Dick, 'Letter To The Editor', *Deep Sky Magazine* #30, Spring 1990, p. 6.
   Hopkins, Stephen, 'Letter To The Editor', *Deep Sky Magazine* #30, Spring 1990, p. 6.
   Bunge, Bob, 'Letter To The Editor', *Deep Sky Magazine* #33, Winter 1990/1, p. 6.

Messier Marathon Logsheet					
Date	Location	Instrument	Number of Objects seen	Objects missed	Notes

4 Other Marathons

This happens every year. It is Marathon morning. You have observed the maximum number of Messier Objects from your location. You ask: 'What's next?' You could wait until the following year and work the Messier Marathon again. Many amateur astronomers do this, and find it becomes easier each year. You can also wait until October or November and see nearly all the Messier Objects in one night (see Chapter 3). You can also go out the next night and do the Messier Marathon all over again with some variation.

Ideas for different Messier Marathons

Photographic or CCD Marathon

How about photographing or obtaining a CCD image of each object? You could image about as many objects as you can see.[1,2]

Photographs for most objects would require fast film and one or two minutes exposure per photograph. Due to the way the Objects are grouped, the 110 Messier Objects would require about ninety-one photographs, if your field of view were three degrees. You can photograph through the telescope or mount the camera piggyback on a guiding scope and use a telephoto lens.

A CCD, or *charged coupled device*, is an 'electronic' camera attached to your telescope. The image is transformed into electronic impulses that are sent to your computer. Many amateurs now own CCDs and specialize in imaging deep sky objects. One advantage of these instruments is that by subtracting the background twilight, they can image when the object is beyond visual reach. Generally speaking, the images can be captured in just a few seconds and examined immediately.

Unaided-eye Marathon

An unaided-eye Messier Marathon could pick up several dozen objects. Table 2.1 shows us that ten Objects are brighter than magnitude 5.0, while twenty are brighter than magnitude 6.0. Why not try for those 'bright' Messier Objects considered being brighter than magnitude 7.0? That would be forty-one Objects. Darkened tubes held to the eyes (lensless binoculars) may help increase contrast.

Small telescope or binocular Marathon

Messier used a small refractor to fill most of his Catalogue. How about pulling the 60 mm refractor out of the closet and trying to do what he did? At this aperture it will be difficult to see the half-dozen Objects near the horizon, but the other 104 or so targets will be less challenging.

Most telescopes have a finder. How many Messier Objects can you find in your finder?

Binoculars are popular at Messier Marathons. Small or large apertures can be used to sail from one place to another.

Bordering on the bizarre

One twist to the Marathon idea is to observe the Messier Objects in numerical order. Brent Archinal saw M2–M34 in one hour on July 30, 1981.

If you live too far north or south to see 110 Messier Objects in one night, what is the shortest span of time over which you can see the whole Catalogue? Begin counting on the last night you can see the setting Objects in the western sky at evening twilight (during March this would be M77, M74, M33), and stop counting days when you can see the 110th Object rising in the eastern sky before morning twilight. Sometime during those days go out and find all the other Objects in between. In my early years of doing the Messier Marathon, from 37 degrees north latitude I last saw M74 on March 19 and had my first sighting of M30 on the morning of March 31, giving me a span of twelve days. With better horizons I could have shortened that span.

Other ideas might include

- City Messier Marathon: Do you want a challenge? You have a lot going against you: lights and high horizons, to name a few. But you are likely to have a lot of guests.
- My Own Back Yard Marathon: This book suggests finding an 'ideal' location for the Messier Marathon. That means re-locating for the night, not always easy for everyone. Why not just go into your backyard and see what you can rack up there? You may not get the high numbers, but you'll get home cooking!
- Star Party Messier Marathon: I'm not talking about the Astro-Club meeting at an ideal site, but you and a tiny group. You are the tour guide. You supply the telescope, the sky supplies the Messier Objects. You may be the only one to stay up the whole night.
- Planet Marathon Too: Whether or not you consider there to be eight or nine planets, try tying in the major planets with a Messier Marathon. Not all planets are visible each year, so this will need some advance planning.

Mini Messier Marathon

Here is a great idea if you don't like staying up all night, you wish to take your time examining the Messier Objects, or your weather is unpredictable. Do several mini Messier Marathons throughout the year, finding twenty-five to thirty Objects each night. Or do only two half-nighters, spaced several months apart, and find fifty to sixty Objects each night. This is easy enough to repeat annually. By doing this you will see every Messier Object each year.

Messier-plus Marathon

This was developed in 1981 by the Saguaro Astronomy Club of Phoenix, Arizona. It is designed to take place in September of each year to provide a second 110-Object Marathon. Seventy-five Messier Objects and thirty-five non-Messier Objects are compiled into an observing order. The Messier Objects that are difficult to see during that month because they are near the sun (these are mainly the Virgo galaxies) are replaced by the non-Messier Objects. They are then assembled into an observing order. You can work through the Saguaro Astronomy Club's list,[3] or make up your own.

Make your own Marathon

If Charles Messier can do it, why can't you? What are your favorite objects? You can make your own list of your best nebulous objects, or a list of your favorite galaxies, or planetary nebulae, or diffuse nebulae, or dark nebulae, or globular or open clusters. It can number a few, or very many. Then pick a night to try to observe most or all of them. Some of you will be happy with seeing the same few dozen objects on your special night each year.

Massive Marathon

As an addition to the Messier Marathon, I once developed my own Massive Marathon. In 1981 I assembled a list of 548 objects. Adding Messier's Catalogue brought this to 658 galaxies, clusters and nebulae. My southern cutoff was −40 degrees. All the objects were previously seen by me with my 10 inch (25 cm) reflector before being added to the list. A Massive Marathon followed. After extensive planning and some practice, on April 3/4, 1981, I located and identified 599 deep-sky objects in one night. My 10 inch reflector was mounted equatorially on a pipe mount, with no setting circles and no clock drive. An average of one object every fifty-two seconds may sound impossible, but I made up a lot of time on the Virgo galaxies. A few months later it took me only a couple of hours to finish up the list. You may wish to try a Massive Marathon too. It is not really as difficult as it might seem; the key is to know the sky and know your telescope.

Those with electronic setting circles (called GOTO telescopes because they 'Go To' wherever you want) can see and identify 1000 or more objects in one night.

Herschel Marathon

As Messier was finishing up his Catalogue, William Herschel of Great Britain cataloged 2477 additional objects. An average of one Marathon per month, covering one to three hours of right ascension and eighty objects each, could group 1000 objects per year.

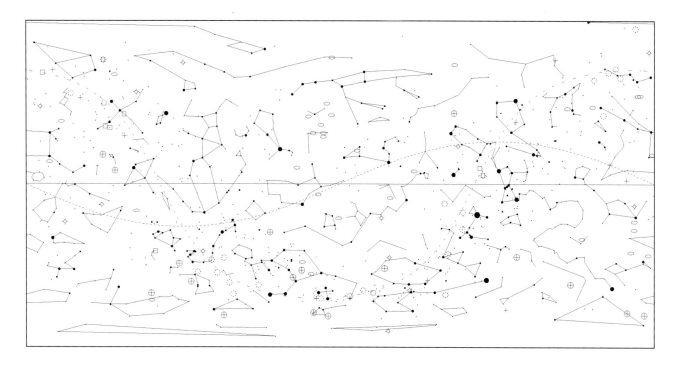

Chart 4.1

A map of the whole sky showing all 109 of the Caldwell objects.

Another idea is to use The Astronomical League's 'Herschel I' list of 400 Herschel objects and go through that in two or more sessions. Even more ambitious would be to use their 'Herschel II' list of 400 even fainter (magnitude 11–13) Herschel objects.[4]

Caldwell Marathon

The Caldwell Catalogue is a list of 109 deep-sky gems compiled in 1995 by the British astronomer, Patrick Moore. It includes none of the Messier Objects. These objects span nearly the whole sky, from 85 degrees north to 80 degrees south.

The Caldwell objects are shown in Chart 4.1. The equator and ecliptic are shown, as is the plane of the Milky Way. Symbols are the same as on previous maps, with a plus sign '+' used for non-NGC objects.

Because of the great north–south span of Caldwell objects, one would have to be near the equator to see all 109 objects from one location. Seeing all the objects in one night would not be possible; the objects are scattered around the celestial sphere much more uniformly than the Messier Objects.

Is there a time where a northern observer and a southern observer, possibly twins, could observe the sky on the same night from equal but opposite latitudes, and see all the objects? First, one would have

Chart 4.2

A map of the Caldwell (Half-)Marathon on November 25 from 35° north latitude.

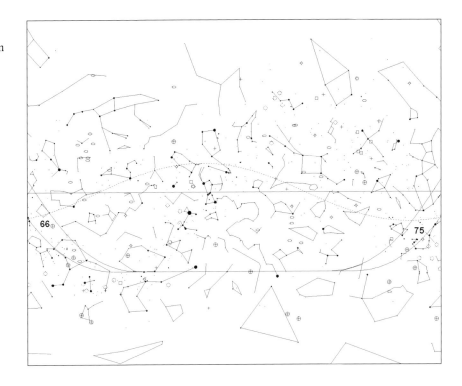

to find the best date, determined by when the sun would interfere with the least number of Caldwell objects. The best opportunity seems to be when the sun is in Scorpius, on or near the evening of November 25.

Someone at 35 degrees north would see the objects shown between the horizon lines in Chart 4.2, while the other at 35 degrees south would see those objects between the horizon lines on Chart 4.3.

A better idea would be to see the objects six months apart, once from each hemisphere. As long as you pick two nights five to seven months apart, and you are at least as far as 30 degrees from the equator, you will be able to see all the objects, with ample overlap.

Lacaille Marathon

Abbe Nicolas Louis de la Caille (Lacaille), who lived in France near the time of Messier, listed forty-two objects in the southern hemisphere. They consisted of three groups of fourteen objects, each group listed by right ascension. The first group is known as 'I1' to 'I14', the second is 'II1' to 'II14', and the third is 'III1' to 'III14'. Of these forty-two, five are non-existent, leaving thirty-seven known Lacaille objects.[5]

Chart 4.3

A map of the Caldwell (Half-)Marathon on November 25 from 35° south latitude.

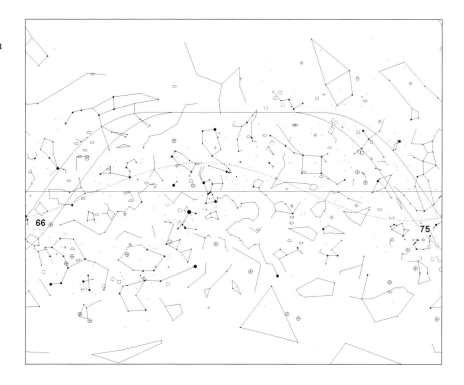

Chart 4.4 is a map of all the known Lacaille objects. The equator and ecliptic are shown, as is the plane of the Milky Way. You may notice that most of the objects are located near the Milky Way. Each object is indicated by a '+', and is identified by Lacaille's Catalogue number.

Since these objects are south of the equator, it is easier to see them from there than from the northern hemisphere. In fact, between January 28 and October 5, a span of eight months, all thirty-seven of the Lacaille objects can be seen in one night from any latitude further south than 20 degrees south latitude. The sky on May 1 is shown in Chart 4.5.

As we move further north the observing season for the Lacaille objects shortens. From 10 degrees south latitude one can see all thirty-seven objects between January 30 and July 3. A map showing the sky from that latitude for May 1 is seen on Chart 4.6.

From the equator the window is even shorter, lasting only six weeks. It extends from April 15 to May 27. This is demonstrated on Chart 4.7.

At 10 degrees north latitude the best we can do is to miss one

Chart 4.4

The Lacaille Catalogue.

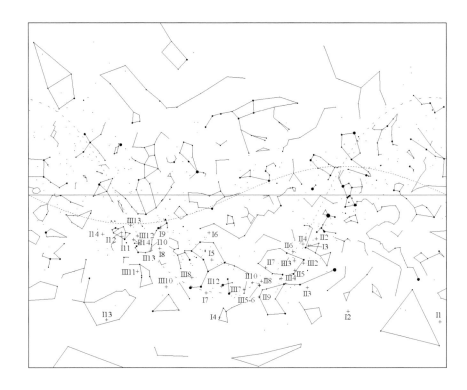

Chart 4.5

The Lacaille objects seen from 20° south latitude on May 1.

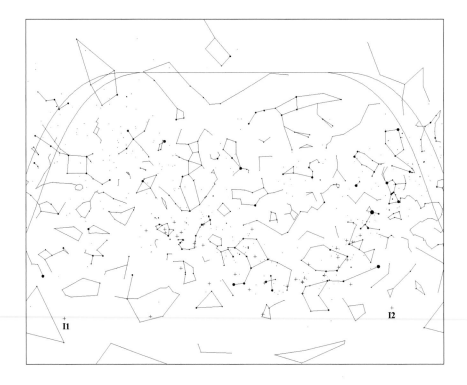

Chart 4.6

The Lacaille objects seen from 10° south latitude on May 1.

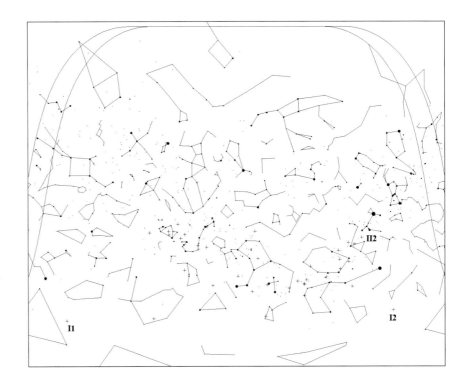

Chart 4.7

The Lacaille objects seen from the equator on May 1.

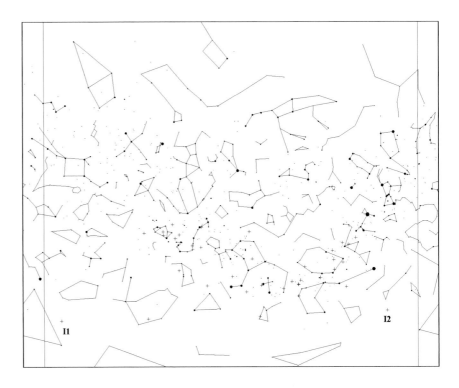

Chart 4.8

The Lacaille objects seen from 10° north latitude on April 1.

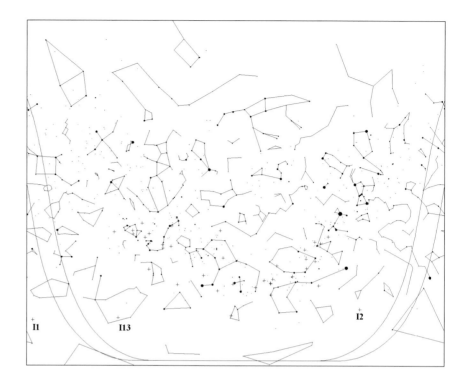

Chart 4.9

The Lacaille objects seen from 20° north latitude on March 1.

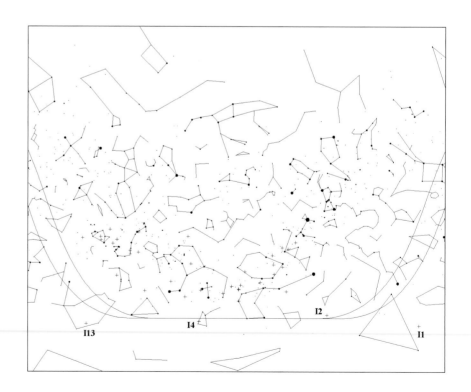

object: that is between March 1 and April 25. The one object we don't get (similar to M30 in the Messier Marathon) is I1, also known as NGC 104, 47 Tucanae.

At 20 degrees north latitude one can see all but three of the Lacaille objects between late February and late March. This is shown in Chart 4.9.

As one travels further north it becomes more difficult to see the Lacaille objects, and the Messier Marathon becomes more attractive.

Notes

1 Raymo, Chet, 'Shootout at Star Hill Inn', *Sky and Telescope*, (Vol. 82), October 1991, pp. 366–370).
2 'Photographic Messier Marathon', *Sky and Telescope*, (Vol 77), March 1989, p. 338.
3 A list of these objects is available at this web site: http://seds.lpl.arizona.edu/messier/xtra/similar/mm_plus.html
4 A list of their observing clubs can found at: http://www.astroleague.org/al/obsclubs/obsclub.html. You can purchase the Herschel books from: Astronomical League Sales, PO Box 572, West Burlington, Iowa 52655, USA. The Herschel Club chairperson is Brenda Branchett, 515 Glen Haven Drive, Deltona, Florida 32738, USA.
5 Good reading about the Lacaille Catalogue can be found at http://seds.lpl.arizona.edu/messier/xtra/history/lacaille.html.

Appendix: The Greek alphabet

The letters of the Greek alphabet are used to identify stars, which are labeled in order of brightness (more or less) for each constellation.

alpha	α	eta	η	nu	ν	tau	τ
beta	β	theta	θ	xi	ξ	upsilon	υ
gamma	γ	iota	ι	omicron	o	phi	φ
delta	δ	kappa	κ	pi	π	chi	χ
epsilon	ε	lambda	λ	rho	ρ	psi	ψ
zeta	ζ	mu	μ	sigma	σ	omega	ω

Glossary

altazimuth mount a binocular or telescope mounting designed to move horizontally (azimuth) and vertically (altitude). The axes are similar to those of a camera tripod.

aperture the diameter of the main objective of a telescope or binoculars.

arcminute one-sixtieth of a degree.

astronomical twilight the time when the sun is 18 degrees below the horizon. The sky is then considered dark enough for serious observing.

comet a 'dirty snowball' in orbit around the sun. It consists of a nucleus (about 3–10 miles across), the head (or coma), and the tail.

constellation a group of stars. There are eighty-eight recognized constellations.

declination a part of the sky coordinate system, similar to latitude on the earth. It runs both north (+) and south (−) of the equator, with 0 degrees at the equator and +/− 90 degrees at the poles.

degree a unit of angular measurement. A full circle around the sky is 360 degrees. The moon subtends ½ degree.

diffuse nebula a mass of dust and gas, illuminated and/or excited by nearby stars.

Dobsonian the name for an altazimuth mount built low to the ground consisting of a 'rocker box' which rotates and holds the telescope tube.

equatorial mount a telescope mount pointed such that a polar axis is aligned with the celestial pole. The axes allow the telescope to move parallel to right ascension and declination. Motion around only one axis, the polar axis, will follow the diurnal motion of the stars.

galaxy a large collection of billions of stars, gas and dust.

globular cluster a tightly compressed collection of many stars, generally located near the hub of our galaxy.

magnitude the brightness of a heavenly body in numerical form. The brighter the object, the lesser the number. Each magnitude is roughly 2.5 times brighter than the next magnitude. The unaided eye can see to magnitude 6, the North Star is magnitude 2.3, and Sirius, the brightest star in the night sky, is magnitude −1.46.

NGC New General Catalogue of about 8000 non-stellar objects, compiled by Dreyer in 1888.

occultation the passage of one heavenly body behind another so that it disappears from view.

open cluster a loose association of stars moving through space together. They are within our galaxy, generally within a few thousand light years of us.

opposition when an astronomical body is opposite the sun in the sky; it is best placed at local midnight.

planetary nebula a mass of gas, ejected by a star, generally in a round shape.

reflector a telescope which gathers light by means of a mirror.

refractor a telescope which gathers light by means of a lens.

right ascension part of the sky coordinate system, similar to longitude on earth. It is measured eastward in hours (twenty-four in all) and minutes (sixty to an hour). For example, 15:24 means 15 hours, 24 minutes east of the March equinox point.

Schmidt–Cassegrain a telescope using both a front lens and a back mirror to gather light.

transit the passage of one heavenly body in front of another.

Bibliography and further reading

Magazine articles

Archinal, Brent, 'The Messier Marathon', *Deep Sky Monthly* #58, pp. 4–9.
   A thorough article covering the Messier Marathon, with a search sequence.
Freeman, David, 'Extreme Stargazing', *The Atlantic Monthly*, March 2000, pp.
      105–107.
   A visitor writes of his night at the All-Arizona Messier Marathon.
Harrington, Phil, 'A Messier Marathon', *Sky and Telescope*, January 1985, pp. 81–82.
   Phil describes the Marathon, providing hints for completing it.
Harrington, Phil, Letters to the Editor, *Deep Sky Magazine* # 36, Autumn 1991, pp.
      10–11.
   Phil recounts results by him and others of the 1991 Messier Marathon.
Harrington, Phil, 'Running the Celestial Marathon', *Astronomy Magazine*, March
      1994, pp. 61–65.
   A complete article on the Messier Marathon, with a search sequence.
Hoffelder, Tom, Letters to the Editor, *Sky and Telescope*, May 1985, p. 388.
   Tom relates the right-angle sweep method of finding the Objects.
Hoffelder, Tom and Hoffelder, Lynn, 'The Messier Marathon – A One Night
      Stand', *Deep Sky Magazine* # 26, Spring 1989, pp. 28–9.
   A good overview with a search sequence using right-angle sweeps. Six minor
      corrections are presented by Tom Hoffelder in the Summer 1989 issue of the
      same magazine, p. 7.
Houston, Walter Scott, 'Deep-Sky Wonders', *Sky and Telescope*, Sept. 1988, p. 323.

Bibliography and further reading

Walter discusses the March 1988 Messier Marathon by Florida amateurs.

Houston, Walter Scott, 'Deep-Sky Wonders', *Sky and Telescope*, June 1991, p. 670.
    Walter relates Harvard Pennington's plans to write a book on the Messier
        Marathon.

Lorenzin, Tom, Letter To The Editor, *Deep Sky Magazine* # 23, Summer 1988, p.8.

Lorenzin, Tom, 'Winning the Messier Marathon', *Astronomy Magazine*, Nov. 1988,
        pp. 120–121.
    Tom relates his experiences concerning the March 19/20, 1988, Messier
        Marathon held by the Charlotte Amateur Astronomers.

Machholz, Don, 'Notes on a Messier Marathon'. Astro-Northwest Convention
        Proceedings, Portland Astronomical Society, Portland, Oregon, 1979.
    Taken from a speech given at the convention.

Machholz, Don, 'The Massive Marathon', *Deep Sky Monthly* # 58, p. 10.
    A short summary of a Massive Marathon activity in 1981.

Reynolds, Jay Freeman, 'Messier Surveys by the Dozen', *Sky and Telescope*, March
        2001, pp. 108–112.
    An expert amateur astronomer describes his observations of the Messier
        Objects with various instruments.

Unknown author, 'Messier Marathoning in Illinois', *Astronomy Magazine*, August
        1994, p. 91.
    Jeff Copeland and Ed Neuzil relate their experiences of running the Marathon
        on March 15/16, 1994.

Books

Busler, William, *An Amateur's Guide To the Messier Objects*, 1967, Memphis
        Astronomical Society, Memphis, Tennessee, 55 pp.
    Descriptions and finder charts for each of the Messier Objects.

Garfinkle, Robert, *Star-Hopping, Your Visa to Viewing the Universe*, 1994, Cambridge
        University Press.
    A full chapter (15) is devoted to the Messier Marathon. It includes star hopping
        instructions to find each Object.

Holyoke, Edward, *OBSERVE: A Guide to the Messier Objects*, 1962, 1966, available from
        Astronomical League Sales, P.O. Box 572, West Burlington, IA. 52655-9998.
    Descriptions of each of the Messier Objects.

Newton, Jack, *Deep Sky Objects, A Photographic Guide for the Amateur*, 1977, Gall
        Publications, 1293 Gerrard St. East, Toronto, Canada, M4L 1 Y8, 160 pp.
    Provides maps and photos of the Messier Objects.

O'Meara, Stephen, *The Messier Objects*, 1998, Cambridge University Press.
    A look at each Messier Object.

Pennington, Harvard, *The Year-Around Messier Marathon Field Guide*, 1997, Willmann-
        Bell, Inc., Richmond, VA.
    A thorough study of the Marathon, with search sequences.

Sarna, Thomas, *Sarna Deep Sky Atlas*, Willmann-Bell, Inc., Richmond, VA.
    Detailed charts of each of the Messier Objects, plus more. White stars on a blue
        background.

Bibliography and further reading

Vehrenberg, Hans, *Atlas of Deep-Sky Splendors*, 1971, Sky Publishing Corp., Cambridge, MA, 220 pp.
  Detailed photographs of each of the Messier Objects, plus more.

Web sites*

http://www.seds.org/messier/Messier.html: a very extensive site about Messier, the Marathon, and other catalogs too.
http://www.astrosurf.com/re/messform.htm: the home of the Messier Object Visibility Program.
http://www.davidpaulgreen.com/tumol.html: large Messier site with a free download for your desktop.
http://jwisn.topcities.com/marathon.htm: a personal look at the Messier Marathon.
http://www.astrosurf.com/re/messier1.html: this site includes Messier history and images.
http://www.hawastsoc.org/deepsky/index.html: an extensive listing of deep-sky catalogs.
http://www.astroimages.com/links.htm: links to dozens of astro-imagers.
http://www.thurs.net/dan/images.html: astro-images by Dan Thurs.
http://www.astropix.com/INDEX.HTM: some of the best color astronomy shots on the web.
http://nav.webring.yahoo.com/hub?ring=techpanastrophot&id=5&hub: a collection of web sites.
http://us.geocities.com/messiermarathon/index.html

Part 2
Atlas

Atlas index of Messier Objects

The Messier Marathon
Search Sequence Atlas

Based on an atlas I compiled in 1982 and have used for every Messier Marathon since, the following pages are designed to walk you through the Messier Objects. The left-hand page shows large portions of the sky to help in orientation and to provide guide stars for those wishing to star-hop to the Messier Objects.

The right side of each page contains the Messier Objects, position (equinox 2000), magnitude and size estimates, and search directions. Knowing the field of view of your telescope, move the instrument in the proper directions for the correct distances. About half of the Messier Objects will be visible in your finder, assuming its aperture is at least 30 mm in diameter and you have reasonably good skies.

Additionally, these small maps show each Object and enough stars to identify it. North is up, and east is to the left. These maps are to various scales, each chosen to simplify finding each Object. On the lower left corner you will often find an arc that has a radius of 3 degrees. In some instances the arc is only of 1 degree radius, this will be identified on the map. Those using setting circles (electronic or manual) will find the right side of each page most useful. Here you will find the Messier Object's number, its coordinates, and the map for final centering onto the Object. The following abbreviations are used in the Atlas:

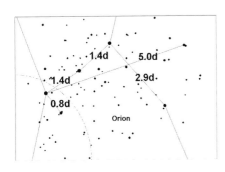

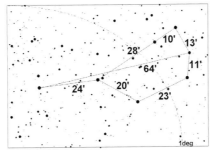

Gal. = galaxy
GL. Cl. = globular cluster
Open Cl. = open cluster
Diff. Neb. = diffuse nebula
Pl. Neb. = planetary nebula

Chart 6.1

Finder size chart. Use this map of the belt of Orion to estimate the size of your finder's field of view. The number of degrees between various stars is shown here. If you are using a Telrad finder, the fields of view should be 4.0 degrees, 2.0 degrees and 0.5 degrees (30 arcminutes).

Chart 6.2

Telescope field size chart. Use the map of the Pleiades (M45) to determine the size of your telescope's field of view. The distances between stars are labeled in arcminutes. Different eyepieces will yield different results. Another method is to turn off your clock drive and measure the amount of time it takes for any star on Chart 6.1 to cross your field of view. Divide the time in minutes by four to calculate the field size in degrees.

**Turn this page to start your
Messier Marathon**

M77

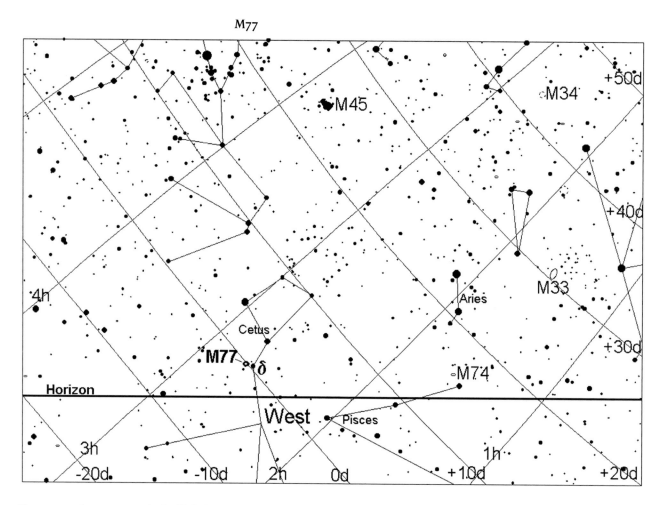

The western sky at astronomical twilight in
mid-March from mid-northern latitudes.

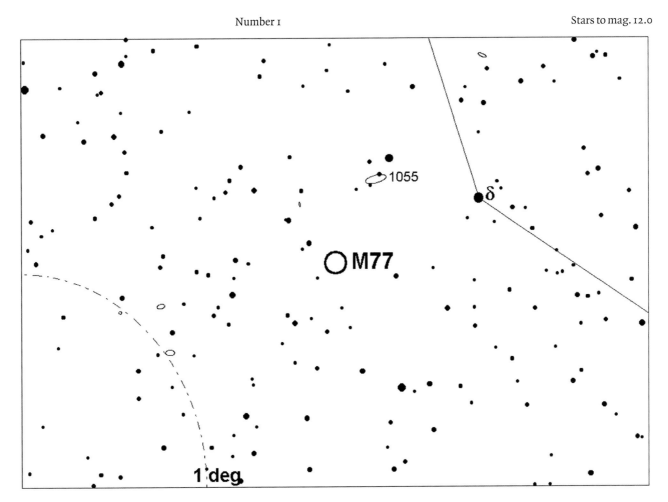

M77 NGC 1068 Gal. 02h 42.7m −00° 01′ mag: 8.9 size: 2′ dia.
From δ Cet. go 0.3° S & 0.9° E to M77.

M74 and M33

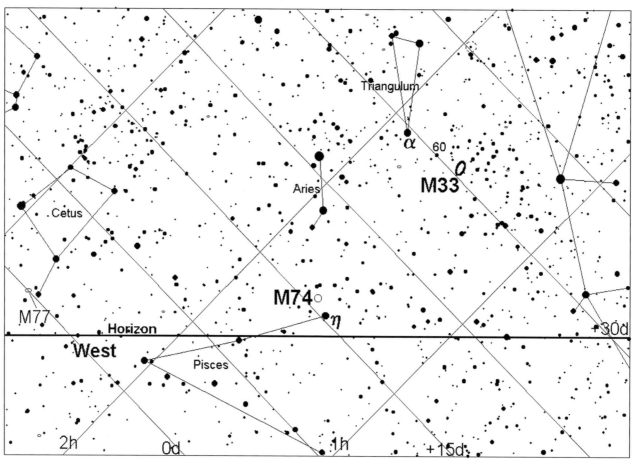

The western sky at astronomical twilight in
mid-March from about +30° latitude.

Number 2 Stars to mag. 11.5

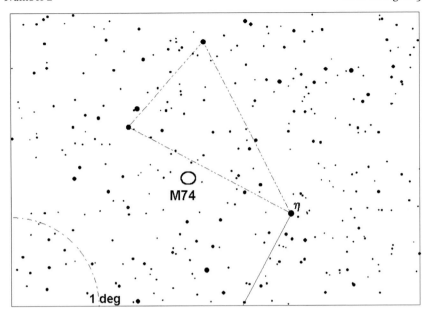

M74 NGC 628 Gal. 01h 36.7m +15° 47′ mag: 9.1 size: 9′ dia.
From η Psc. travel 0.4° N & 1.3° E to M74.

Number 3 Stars to mag. 11.5

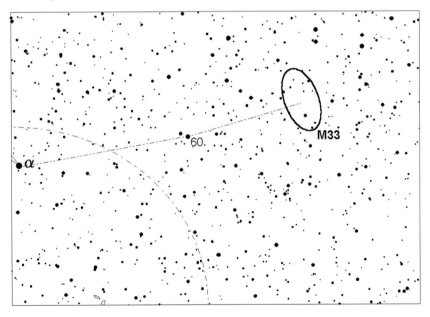

M33 NGC 598 Gal. 01h 33.8m +30° 39′ mag: 6.2 size: 22′ × 16′.
From α Tri. sweep 2.5° WNW to star '60', continue 1.8° to M33.

M31, M32, M110, M52

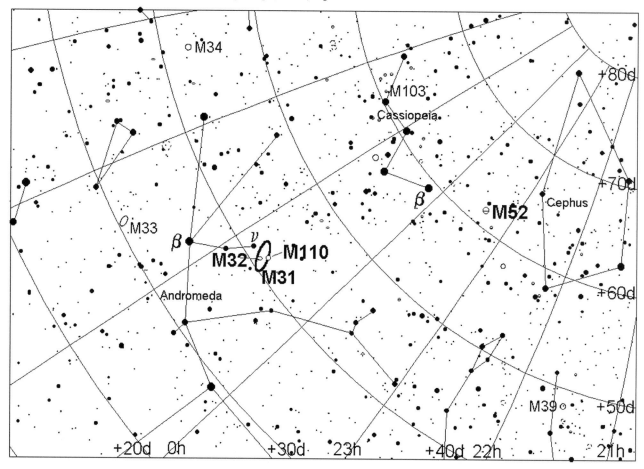

Numbers 4, 5 and 6 · Stars to mag. 9.5

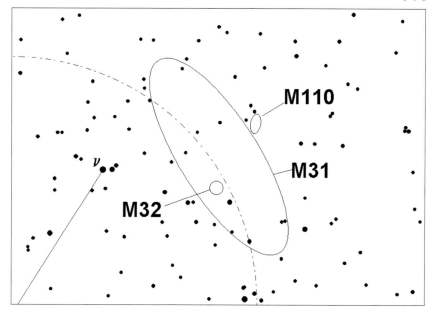

M31 NGC 224 Gal. 00h 42.7m +41° 16′ mag: 4.5 size: 145′ × 20′.
From νAnd. go 1.2° W to M31

M32 NGC 221 Gal. 00h 42.7m +40° 52′ mag: 8.6 size: 2′dia.

M110 NGC 205 Gal. 00h 40.3m +41° 41′ mag: 8.2 size: 10′ × 4′.

Number 7 · Stars to mag. 8.0

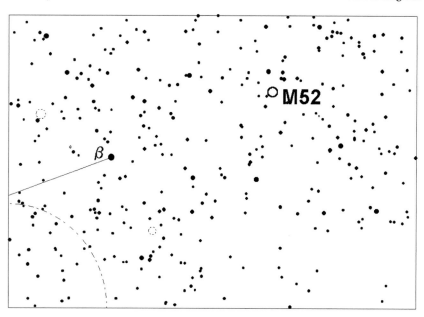

M52 NGC 7654 Open Cl. 23h 24.2m +61° 36′ mag: 7.6 size: 9′ × 7′.
From βCas. go 2.3° N & 5.9° W to M52.

M103, M76, M34 and M45

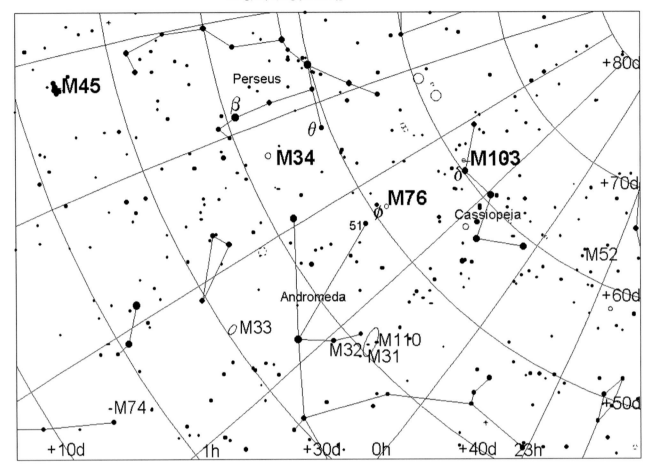

Number 8   Stars to mag. 9.5

Number 9   Stars to mag. 10.0

M103 NGC 581 Open Cl. 01h 33.1m +60° 42′ mag: 7.0 size: 8′ dia. *From δ Cas. go 0.4° N & 1.0° E to M103.*

M76 NGC 650–1 Pl. Neb. 01h 42.2m +51° 44′ mag: 9.6 size: 3′ × 2′. *From φ Per. go 1.0° NNW to M76.*

Number 10   Stars to mag. 8.5

Number 11   Stars to mag. 8.5

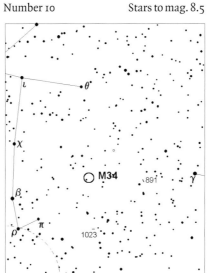

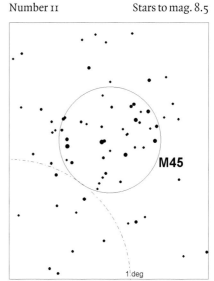

M34 NGC 1039 Open Cl. 02h 42.0m +42° 47′ mag: 6.3 size: 25′ × 15′. *From β Per. go 0.8° N & 4.8° W to M34.*

M45 Mel 22 Open Cl. 03h 47.5m +24° 07′ mag: 2.0 size: 70′ × 40′. *An easy naked-eye object in Taurus.*

M79, M42, M43 and M78

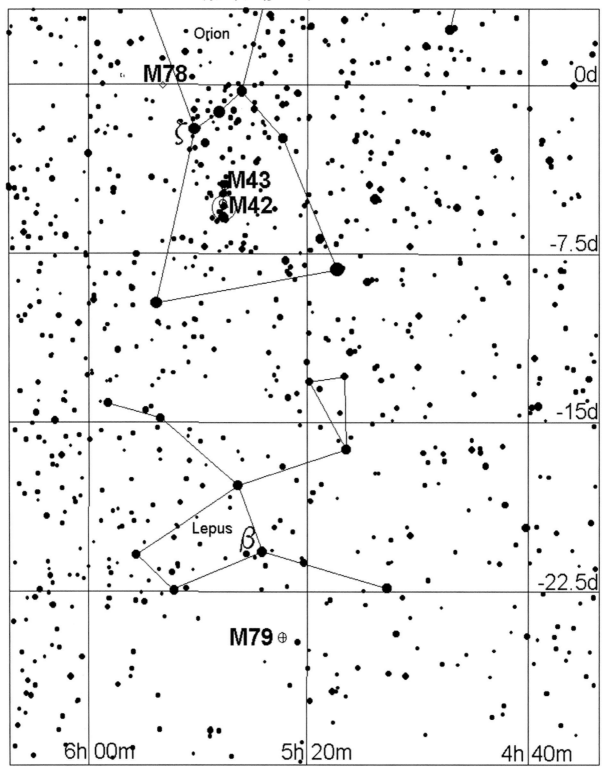

The Messier Marathon Search Sequence Atlas

Number 12   Stars to mag. 8.5   Numbers 13 and 14   Stars to mag. 8.5

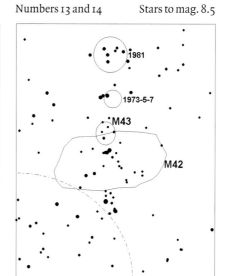

M79 NGC 1904 GL. Cl. 05h 24.2m −24° 31′ mag: 8.1 size: 3′ dia.
From β Lep. go 1.0° W & 3.9° S to M79.

M42 NGC 1976 Diff. Neb. 05h 35.3m −05° 23′ mag: 4.3 size: 40′ × 30′.
This is the middle 'star' in the sword of Orion.

M43 NGC 1982 Diff. Neb. 05h 35.5m −05° 16′ mag: 8.3 size: 5′ dia.
From M42 travel 0.1° N to M43.

Number 15   Stars to mag. 8.0

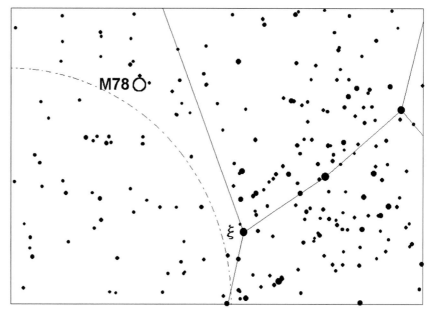

M78 NGC 2068 Diff. Neb. 05h 46.7m +00° 04′ mag: 8.4 size: 8′.
From ζ Ori. go 1.6° E & 2.0° N to M78.

M41, M93, M47 and M46

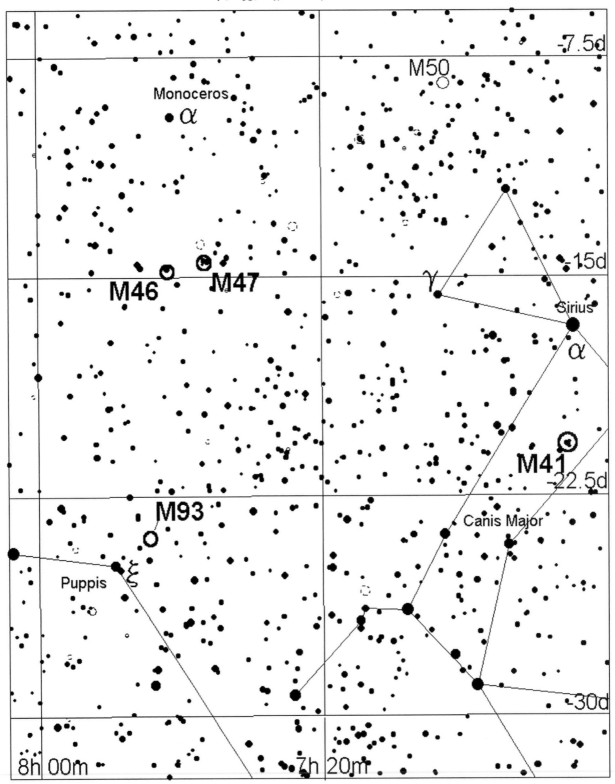

Number 16 Stars to mag. 10.5 Number 17 Stars to mag. 11.0

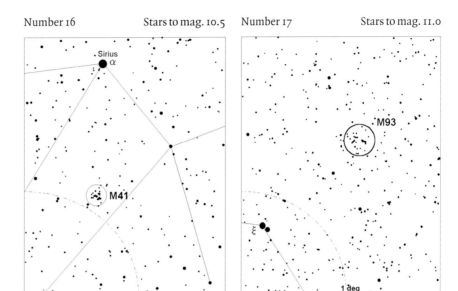

M41 NGC 2287 Open Cl. 06h 47.0m −20° 45′ mag: 5.8 size: 25′ dia. *From Sirius go* 0.3° *E &* 4.0° *S to M41.*

M93 NGC 2447 Open Cl. 07h 44.6m −23° 53′ mag: 5.8 size: 12′ × 8′. *From ξ Pup. go* 1.0° *W &* 1.0° *N to M93.*

Numbers 18 and 19 Stars to mag. 11.0

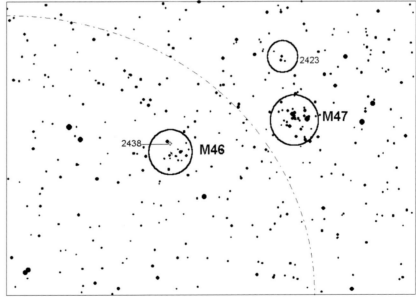

M47 NGC 2422 Open Cl. 07h 36.6m −14° 29′ mag: 5.6 size: 17′ × 12′. *From M93 go* 1.5° *W &* 9.4° *N to M47.*

M46 NGC 2437 Open Cl. 07h 41.8m −14° 49′ mag: 6.7 size: 18′ × 15′. *From M47 go* 0.3° *S &* 1.3° *E to M46.*

The planetary nebula NGC 2438 is in M46.

M50 and M48

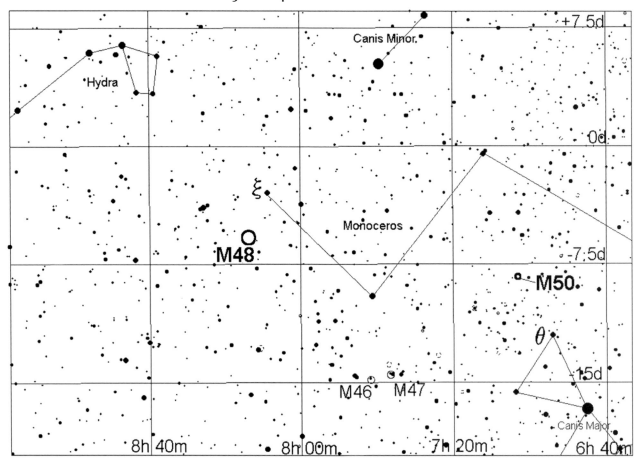

Number 20   Stars to mag. 10.5

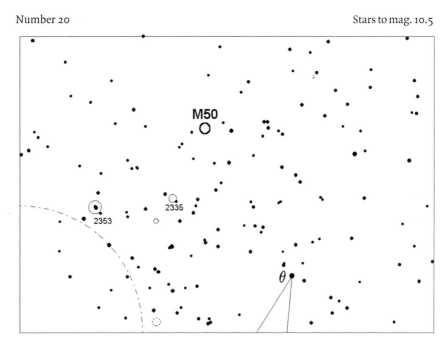

M50 NGC 2323 Open Cl. 07h 03.0m −08° 21′ mag: 6.4 size: 8′ × 6′.
From θCMa. go 2.1° E & 3.7° N to M50.

Number 21   Stars to mag. 8.0

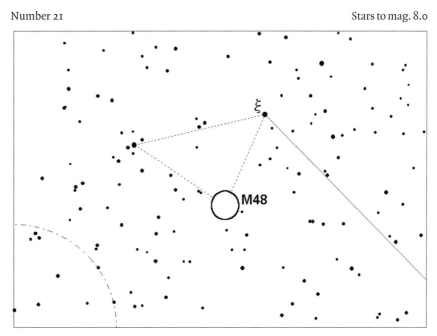

M48 NGC 2548 Open Cl. 08h 13.8m −05° 48′ mag: 6.2 size: 40′ × 35′.
From ξ Mon. go 1.2 E & 2.8° S to M48.

M1, M35, M37, M36 and M38

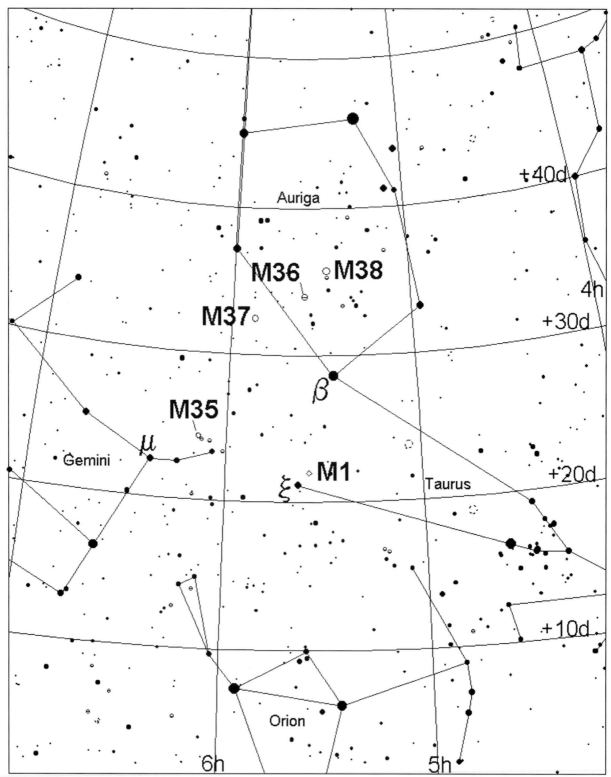

Number 22 — Stars to mag. 9.5

Number 23 — Stars to mag. 9.5

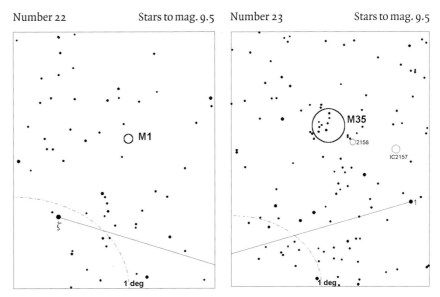

M1 NGC 1952 Diff. Neb. 05h 34.5m +22° 01′ mag: 8.7 size: 7′ × 4′.
From ζ Tau. go 0.7° W & 0.8° N to M1.

M35 NGC 2168 Open Cl. 06h 08.8m +24° 20′ mag: 4.8 size: 25′ × 18′.
From μ Gem. go 1.8° N & 3.2° W to M35.

Number 24 — Stars to mag. 9.0

Numbers 25 and 26 — Stars to mag. 9.0

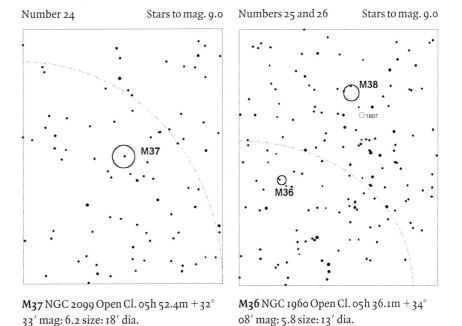

M37 NGC 2099 Open Cl. 05h 52.4m +32° 33′ mag: 6.2 size: 18′ dia.
From β Tau. go 3.9° N & 5.4° E to M37.

M36 NGC 1960 Open Cl. 05h 36.1m +34° 08′ mag: 5.8 size: 13′ dia.
From M37 go 1.7° N & 3.4° W to M36.

M38 NGC 1912 Open Cl. 05h 28.7m +35° 50′ mag: 6.0 size: 11′ dia.
From M36 go 1.6° N & 1.5° W to M38.

M44 and M67

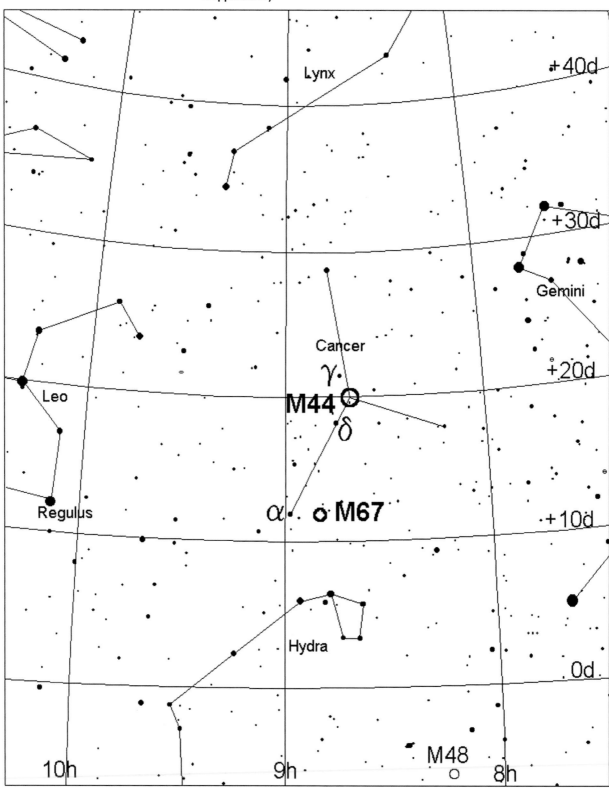

Number 27 Stars to mag. 10.0

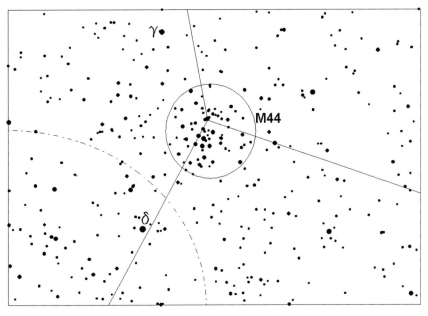

M44 NGC 2632 Open Cl. 08h 40.0m +20° 00′ mag: 3.3 size: 60′ × 50′.
From δ Cnc. go 1.0° W & 1.5° N to M44.

Number 28 Stars to mag. 10.0

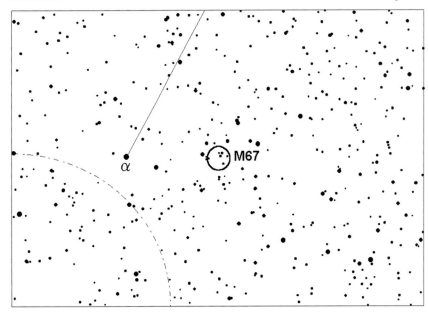

M67 NGC 2682 Open Cl. 08h 51.4m +11° 48′ mag: 7.6 size: 12′ × 10′.
From M44 go 2.4° E & 8.2° S to M67.

M95, M96, M105, M65 and M66

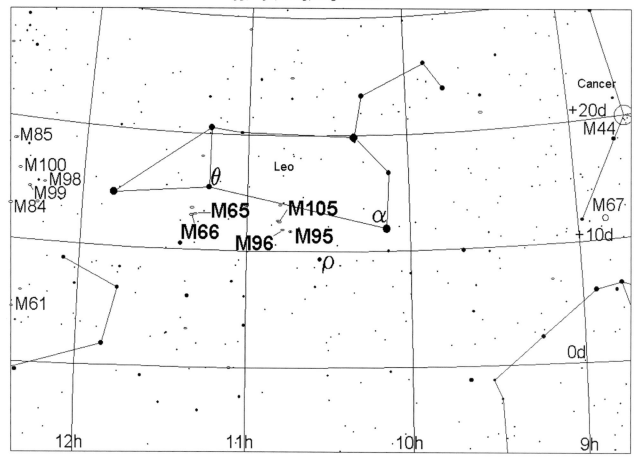

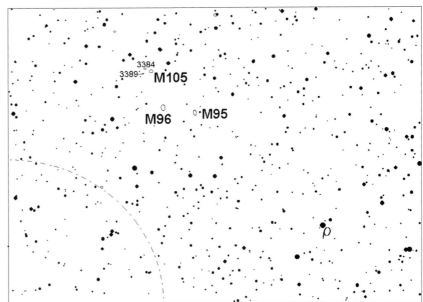

M95 NGC 3351 Gal. 10h 44.0m +11° 42′ mag: 9.5 size: 7′ dia.
From ρ Leo go 2.6° N and 2.7° E to M95.

M96 NGC 3368 Gal. 10h 46.8m +11° 49′ mag: 9.1 size: 8′ × 4′.
From M95 go 0.1° N & 0.7° E to M96.

M105 NGC 3379 Gal. 10h 47.9m +12° 35′ mag: 9.1 size: 4′ dia.
From M96 go 0.8° N & 0.3° E to M105.

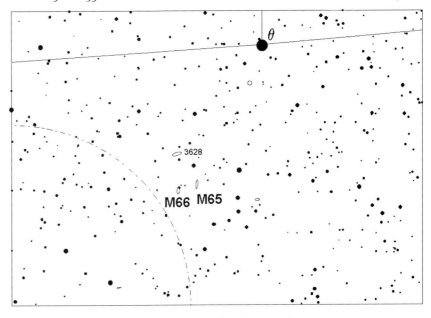

M65 NGC 3623 Gal. 11h 18.9m +13° 06′ mag: 8.9 size: 11′ × 4′.
From θ Leo go 1.8° E & 2.2° S to M65.

M66 NGC 3627 Gal. 11h 20.2m +12° 59′ mag: 8.6 size: 9′ × 5′.
From M65 go 0.1° S & 0.3° E to M66.

121

M81, M82, M97, M108 and M109

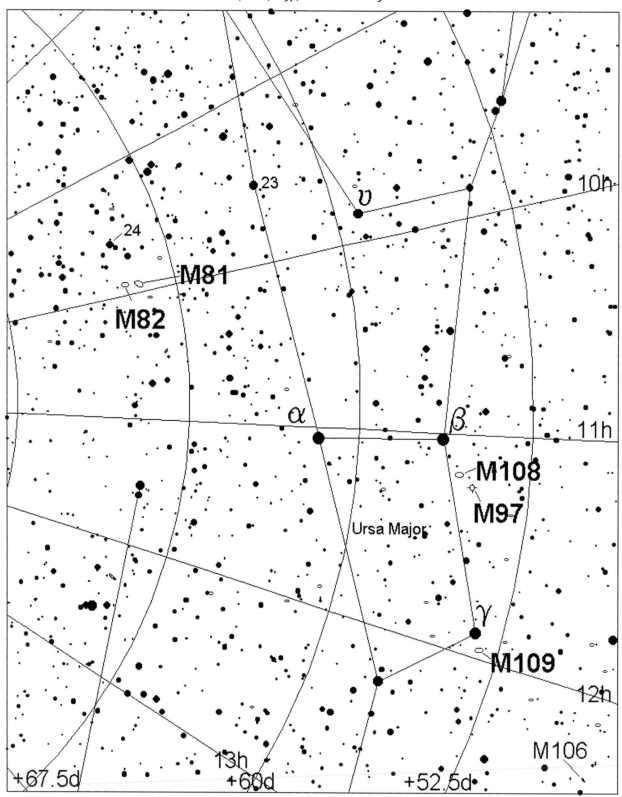

Numbers 34 and 35   Stars to mag. 11.0   Numbers 36 and 37   Stars to mag. 11.0

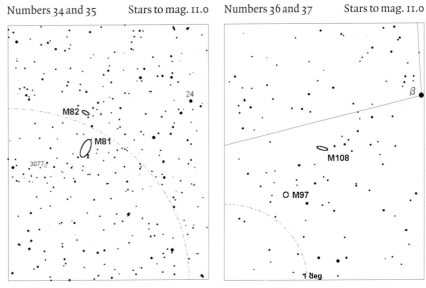

M81 NGC 3031 Gal. 09h 55.6m +69° 04′
mag: 7.3 size: 12′ × 5′.
From '24' UMa. go 0.8° S & 1.8° E to M81.
M82 NGC 3034 Gal. 09h55.8m +69° 44′
mag: 8.3 size: 9′ × 2′.
From M81 go 0.5° N to M82.

M97 NGC 3587 Gal. 11h 14.9m +55° 01′
mag: 9.7 size: 6′ dia.
From β UMa. go 1.4° S & 1.9° E to M97.
M108 NGC 3556 Gal. 11h 11.6m +55° 40′
mag: 9.4 size: 7′ × 2′.
From M97 go 0.5° W & 0.6° N to M108.

Number 38   Stars to mag. 11.0

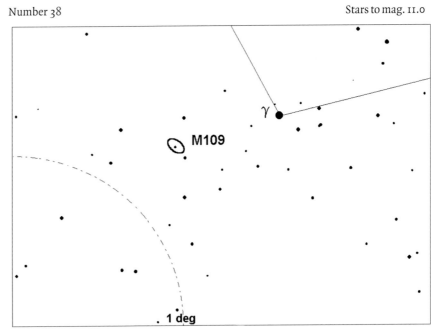

M109 NGC 3992 Gal. 11h 57.7m +53° 22′ mag: 9.6 size: 4′ dia.
From γ UM. go 0.4° S & 0.6° E to M109.

123

M40, M106, M94 and M63

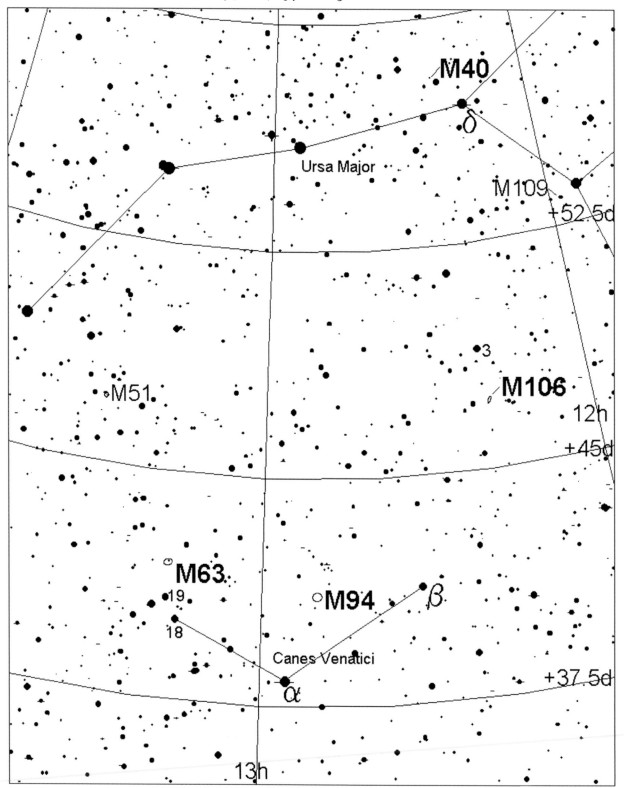

Number 39   Stars to mag. 11.0   Number 40   Stars to mag. 11.0

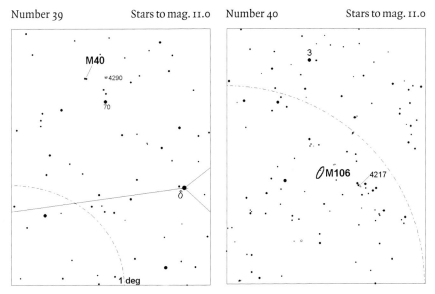

M40 Winnecke 4 Double Star 12h22.2m +58° 05′ mag: 8.7 size: 1′ dia.
From δ U. M. go 1.0° N & 1.0° E to M40.

M106 NGC 4258 Gal. 12h 19.0m +47° 18′ mag: 8.8 size: 11′ × 5′.
From M40 go 0.4° W & 10.8° S to M106.

Number 41   Stars to mag. 11.0   Number 42   Stars to mag. 11.0

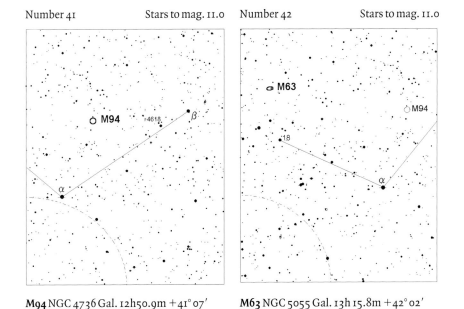

M94 NGC 4736 Gal. 12h50.9m +41° 07′ mag: 8.4 size: 4′ dia.
From α CVn. go 0.9° W & 2.8° N to M94.

M63 NGC 5055 Gal. 13h 15.8m +42° 02′ mag: 8.7 size: 8′ × 4′.
From M94 go 0.9° N & 4.6° E to M63.

125

M51, M101 and M102

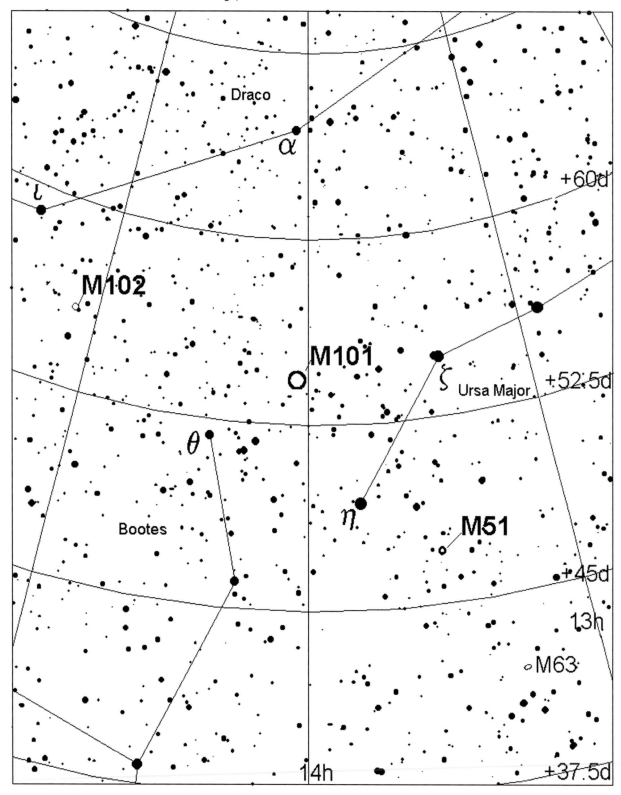

Number 43   Stars to mag. 10.5   Number 44   Stars to mag. 10.5

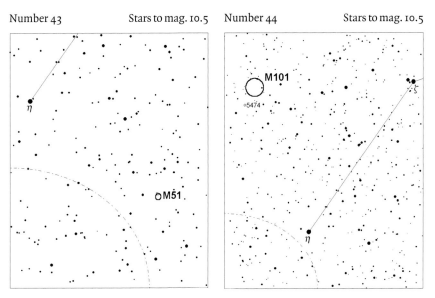

M51 NGC 5194 Gal. 13h29.9m +47° 12′
mag: 7.9 size: 14′ × 11′.
From η UMa. go 2.1° S & 2.9° W to M51.

M101 NGC 5457 Gal. 14h 03.2m +54° 21′
mag: 8.8 size: 14′ dia.
From η UMa. go 2.6° E & 5.0° N to M101.

Number 45   Stars to mag. 10.5

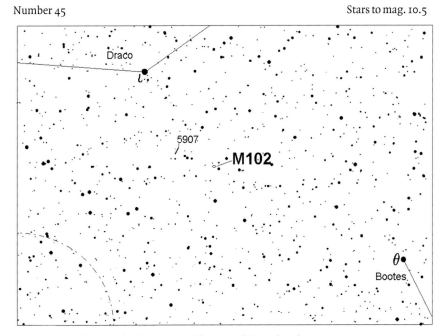

M102 NGC 5866 Gal. 15h 06.5m +55° 46′ mag: 9.6 size: 4′ × 2′.
From ι Dra. go 2.4° W & 3.2° S to M102.

127

M53, M64 and M3

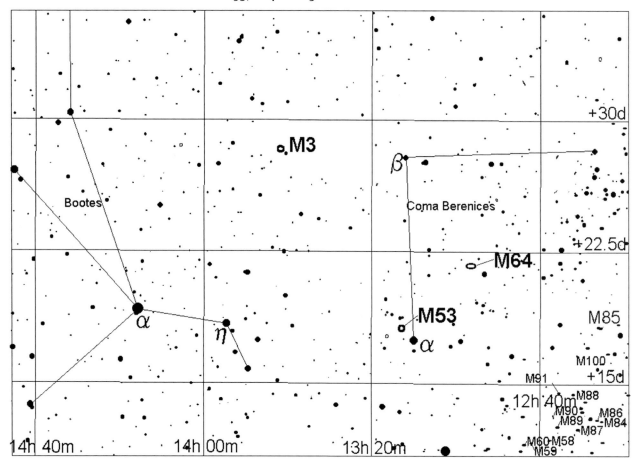

Numbers 46 and 47 Stars to mag. 10.5

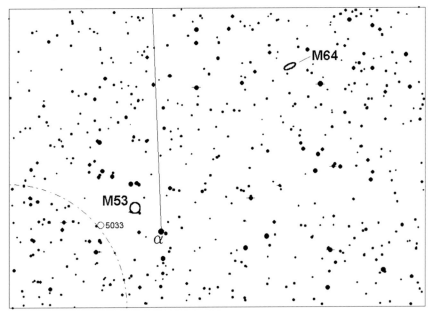

M53 NGC 5024 GL. Cl. 13h 12.9m +18° 10′ mag: 7.4 size: 3′ dia.
From α Com. go 0.6° N & 0.7° E to M53.

M64 NGC 4826 Gal. 12h 56.7m +21° 41′ mag: 8.7 size: 8′ × 4′.
From M53 go 3.5° N & 3.9° W to M64.

Number 48 Stars to mag. 10.5

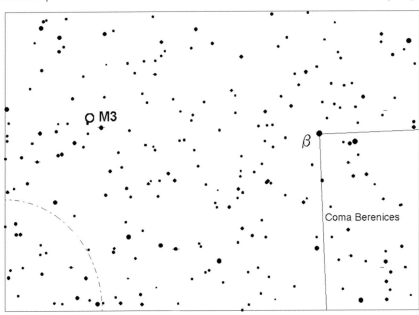

M3 NGC 5272 GL. Cl. 13h42.2m +28° 23′ mag: 6.4 size: 8′ dia.
From β Com. go 0.5° N & 6.7° E to M3.

M98, M99, M100, M85, M84 and M86

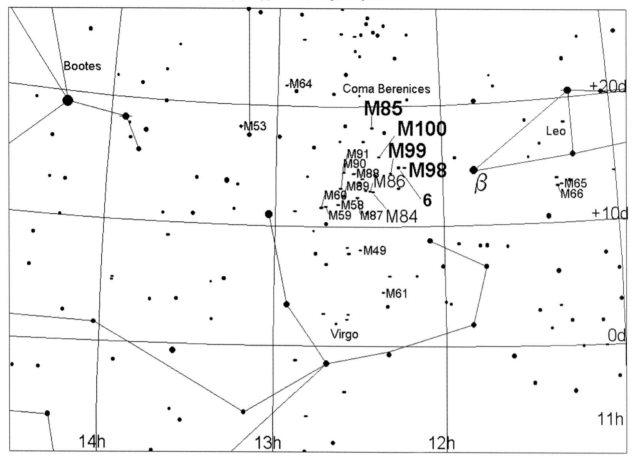

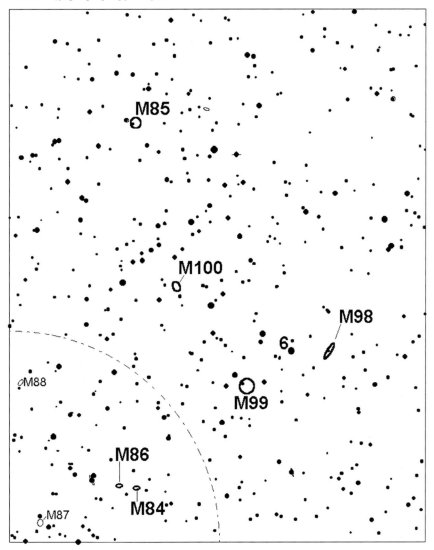

M98 NGC 4192 Gal. 12h13.8m +14° 54′ mag: 9.8 size: 8′ × 3′.
From 6 Com. go 0.5° W to M98.

M99 NGC 4254 Gal. 12h18.8m +14° 25′ mag: 9.4 size: 6′ dia.
From M98 go 0.5° S & 1.2° E to M99.

M100 NGC 4321 Gal. 12h22.9m +15° 49′ mag: 9.6 size: 7′ dia.
From M99 go 1.0° E & 1.4° N to M100.

M85 NGC 4382 Gal. 12h25.4m +18° 11′ mag: 8.9 size: 4′ dia.
From M100 go 0.6° E & 2.4° N to M85.

M84 NGC 4374 Gal. 12h25.1m +12° 53′ mag: 8.8 size: 5′ dia.
From M85 go 5.3° S to M84.

M86 NGC 4406 Gal. 12h26.2m +12° 57′ mag: 8.8 size: 5′ dia.
From M84 go 0.1° N & 0.3° E to M86.

M87, M89, M90, M88, M91, M58, M59 and M60

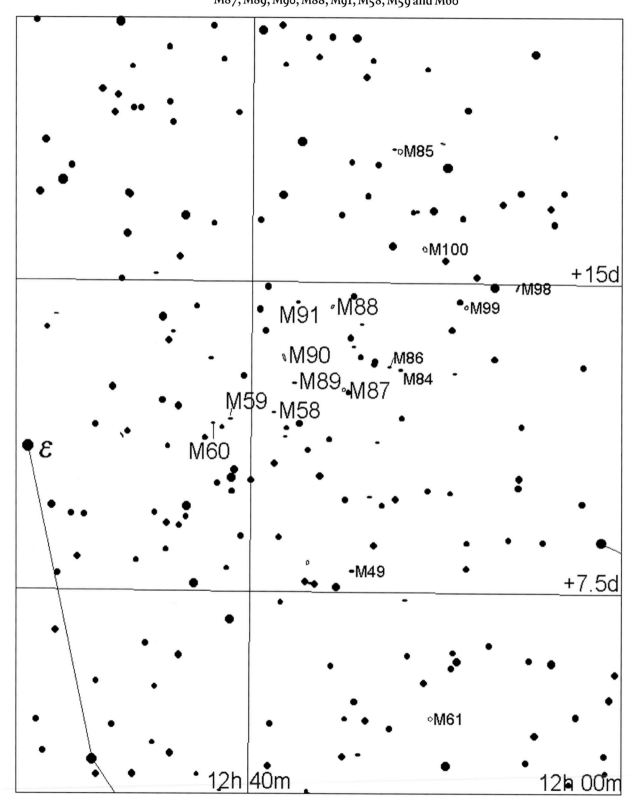

Numbers 55, 56, 57, 58, 59, 60, 61 and 62   Stars to mag. 11.0

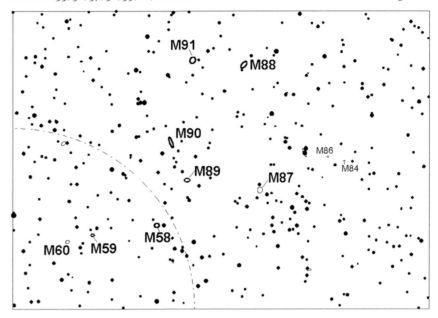

M87 NGC 4486 Gal. 12h30.8m +12° 23′ mag: 8.7 size: 5′ dia.
From M86 go 0.6° S and 1.1° E to M87.

M89 NGC 4552 Gal. 12h35.7m +12° 33′ mag: 9.1 size: 3′ dia.
From M87 go 0.2° N & 1.2° E to M89.

M90 NGC 4569 Gal. 12h36.8m +13° 10′ mag: 9.3 size: 6′ × 2′.
From M89 go 0.3° E & 0.7° N to M90.

M88 NGC 4501 Gal. 12h32.0m +14° 25′ mag: 9.4 size: 8′ × 3′.
From M90 go 1.2° W & 1.2° N to M88.

M91 NGC 4548 Gal. 12h35.4m +14° 30′ mag: 9.9 size: 3′dia.
From M88 go 0.1° N & 0.8° E to M91.

M58 NGC 4579 Gal. 12h37.7m +11° 49′ mag: 9.2 size: 4′ dia.
From M91 go 0.6° E & 2.7° S to M58.

M59 NGC 4621 Gal. 12h42.0m +11° 39′ mag: 9.4 size: 3′ × 2′.
From M58 go 0.2° S & 1.1° E to M59.

M60 NGC 4649 Gal. 12h43.7m +11° 33′ mag: 9.2 size: 4′ × 3′.
From M59 go 0.1° S & 0.4° E to M60.

M49, M61 and M104

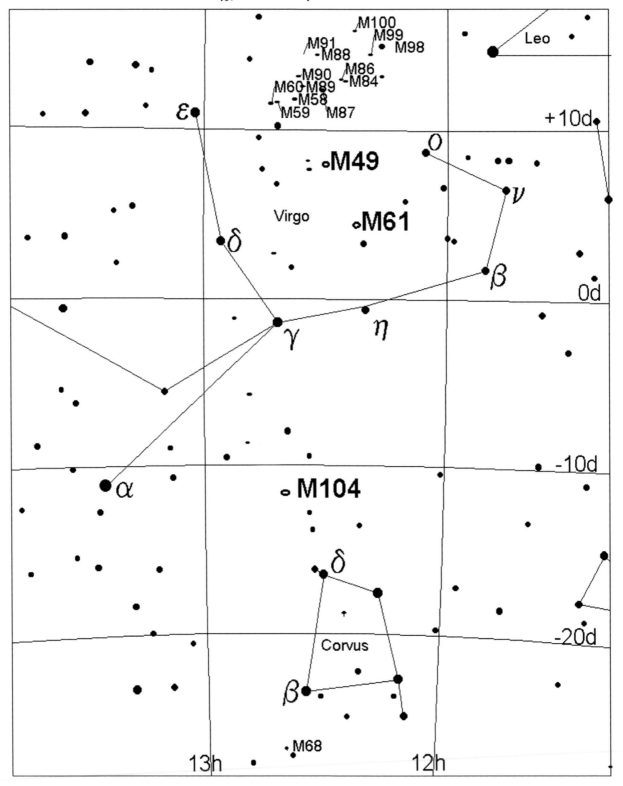

Numbers 63 and 64   Stars to mag. 11.0

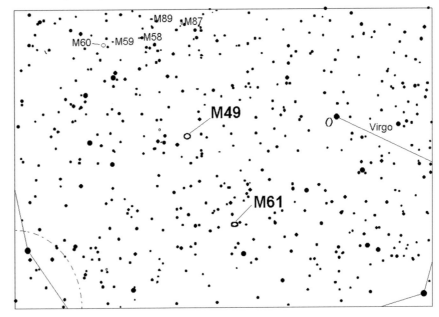

M49 NGC 4472 Gal. 12h29.8m +08° 00′ mag: 8.6 size: 4′ dia.
From M60 go 3.4° W & 3.5° S to M49.

M61 NGC 4303 Gal. 12h21.9m +04° 28′ mag: 9.0 size: 5′ dia.
From M49 go 2.0° W & 3.5° S to M61.

Number 65   Stars to mag. 11.0

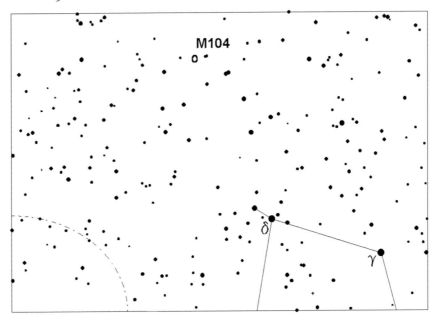

M104 NGC 4594 Gal. 12h40.0m −11° 37′ mag: 8.8 size: 7′ × 4′.
From δ Crv. go 2.4° E & 4.9° N to M104.

M68 and M83

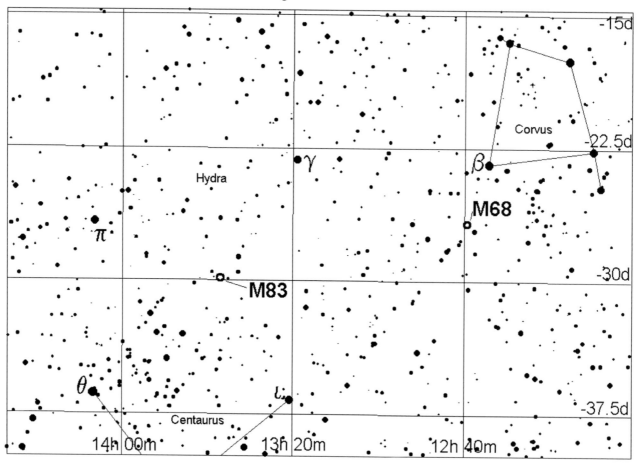

Number 66

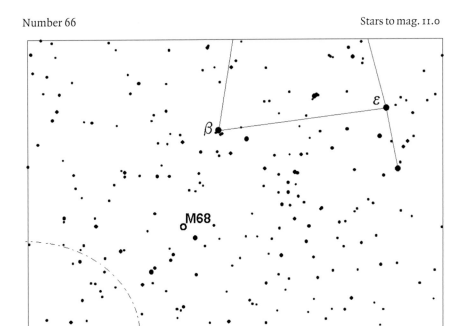

M68 NGC 4590 GL. Cl. 12h39.5m −26° 45′ mag: 8.1 size: 5′ dia.
From β Crv. go 1.1° E & 3.5° S to M68.

Number 67

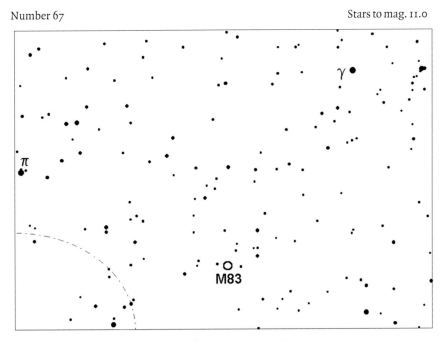

M83 NGC 5236 Gal. 13h37.0m −29° 52′ mag: 7.9 size: 10′ × 7′.
From γ Hya. go 4.2° E & 6.7° S to M83.

M5, M13 and M92

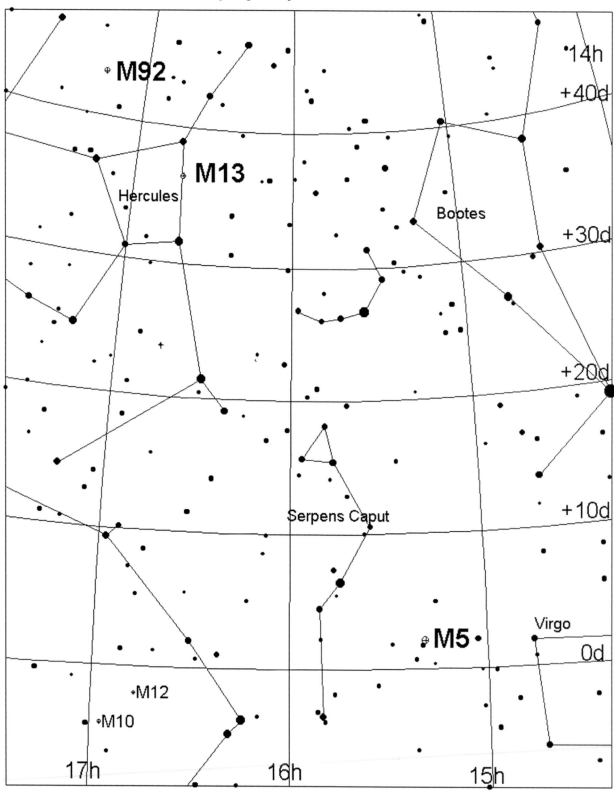

Number 68 Stars to mag. 11.0

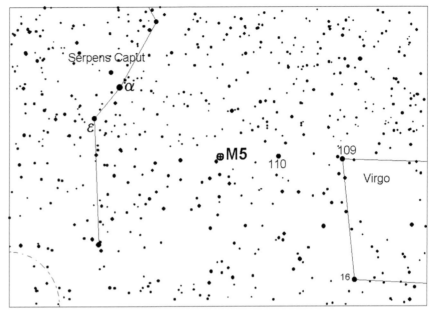

M5 NGC 5904 Gal. 15h18.5m +02° 05′ mag: 6.2 size: 10′ dia.
From 109 Vir. go 0.2° N & 8.1° E to M5.

Numbers 69 and 70 Stars to mag. 11.0

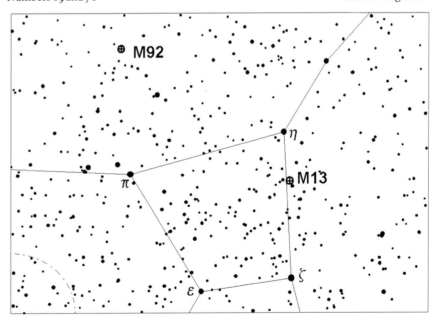

M13 NGC 6205 GL. Cl.16h 41.7m +36° 28′ mag: 6.2 size: 13′ dia.
From η Her. go 0.3° W & 2.5° S to M13.

M92 NGC 6341 GL. Cl. 17h17.1m +43° 08′ mag: 6.9 size: 8′ dia.
From π Her. 0.4° E & 6.1° N to M92.

M57, M56, M29 and M39

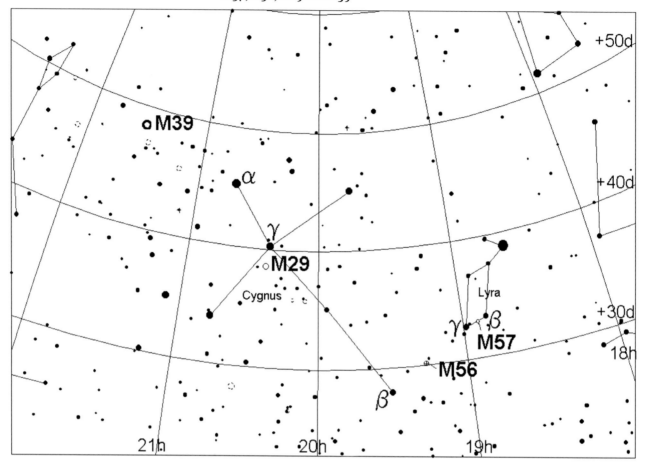

Numbers 71 and 72   Stars to mag. 11.0

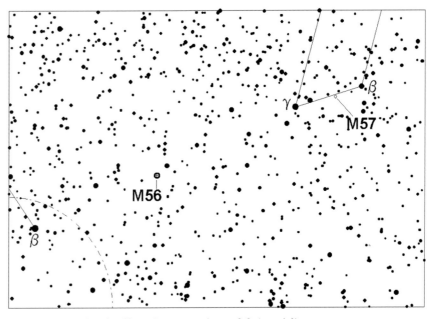

M57 NGC 6720 Pl. Neb. 18h 53.6m +33° 02' mag: 8.8 size: 2' dia.
From β Lyr. go 0.3° S & 0.7° E to M57.

M56 NGC 6779 GL. Cl. 19h16.6m +30° 11' mag: 8.8 size: 5' dia.
From M57 go 2.8° S & 5.0° E to M56.

Number 73   Stars to mag. 11.0   Number 74   Stars to mag. 8.5

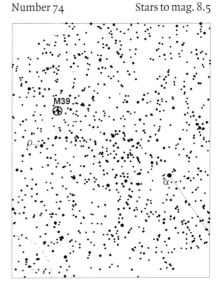

M29 NGC 6913 Open Cl. 20h24.0m +38°
31' mag: 7.4 size: 7' dia.
From γ Cyg. go 0.3° E & 1.7° S to M39

M39 NGC 7092 Open Cl. 21h 32.3m +48°
26' mag: 5.4 size: 30' × 20'.
From α Cyg. go 3.2° N & 9.0° E to M39.

141

M27 and M71

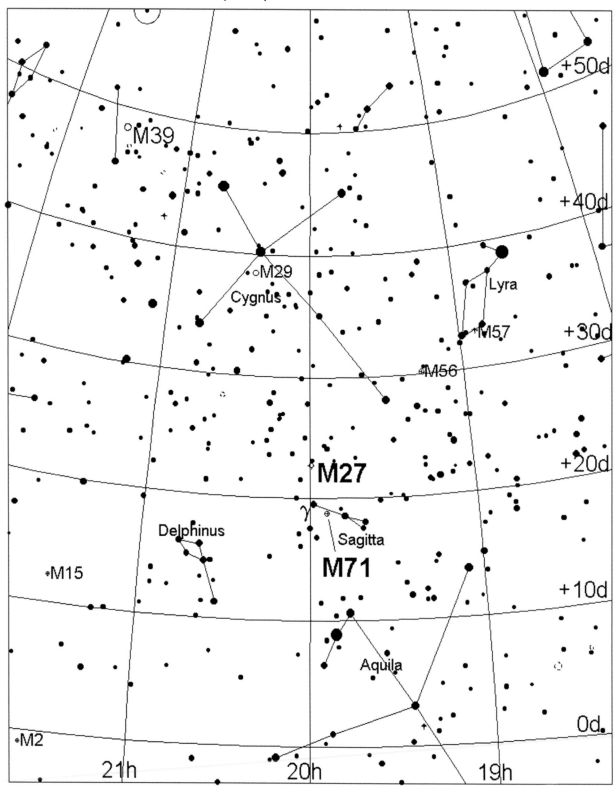

Numbers 75 and 76

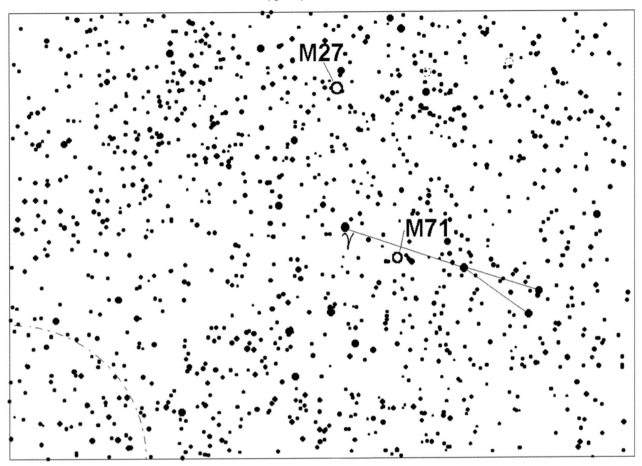

M27 NGC 6853 Pl. Neb. 19h 59.6m +22°
43′ mag: 8.1 size: 7′ dia.
From γ Sge. go 0.2° E & 3.2° N to M27.

M71 NGC 6838 GL. Cl. 19h53.7m +18° 47′
mag: 8.1 size: 7′ dia.
From γ Sge. go 0.7° S & 1.3° W to M71.

M12, M10, M14, M107 and M9

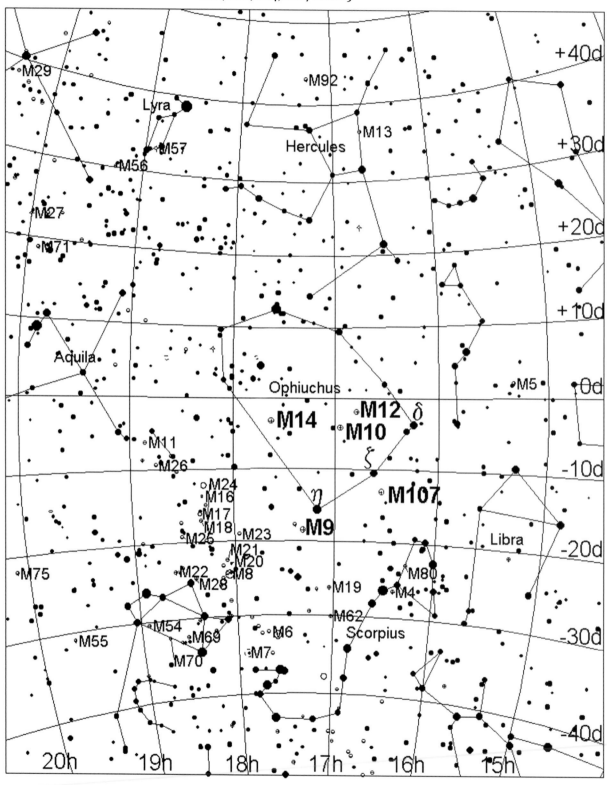

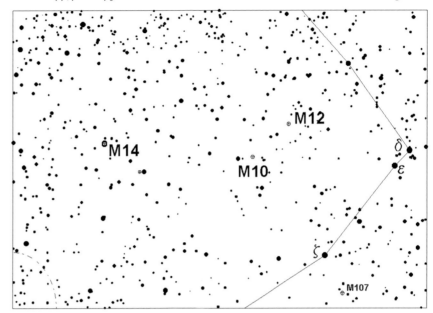

M12 NGC 6218 GL. Cl. 16h 47.2m −01° 57′ mag: 6.8 size: 8′ dia.
From δ Oph. go 1.7° N & 8.2° E to M12.

M10 NGC 6254 GL. Cl. 16h57.2m −04° 06′ mag: 6.7 size: 9′ dia.
From M12 go 2.1° S & 2.5° E to M10.

M14 NGC 6402 GL. Cl. 17h37.6m −03° 15′ mag: 7.7 size: 8′ dia.
From M10 go 0.8° N & 10.1° E to M14.

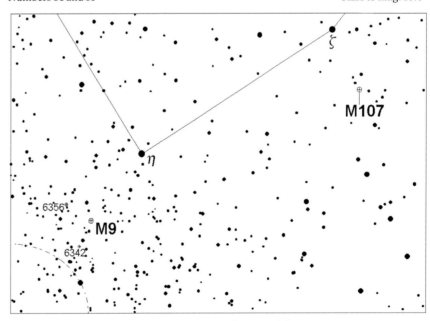

M107 NGC 6171 GL. Cl. 16h32.5m −13° 03′ mag: 8.6 size: 5′ dia.
From ζ Oph. go 1.2° W & 2.5° S to M107.

M9 NGC 6333 GL. Cl. 17h19.2m −18° 31′ mag: 7.3 size: 4′ dia.
From η Oph. go 2.1° E & 2.8° S to M9.

M4, M80, M19, M62, M6 and M7

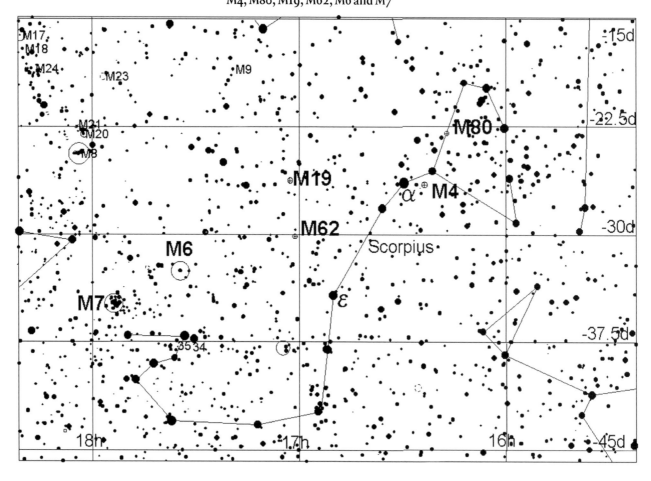

Numbers 82 and 83   Stars to mag. 10.5   Numbers 84 and 85   Stars to mag. 10.5

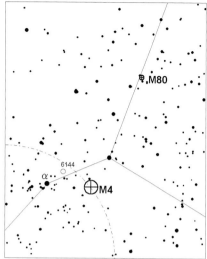
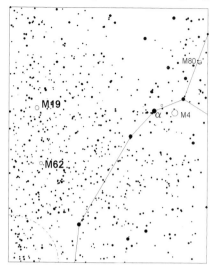

M4 NGC 6121 GL. Cl. 16h23.6m −26° 31′ mag: 6.4 size: 14′ dia.
From α Sco. go 0.1° S & 1.3° W to M4.

M80 NGC 6093 GL. Cl. 16h 17.0m −22° 59′ mag: 8.1 size: 3′ dia.
From M4 go 1.5° W & 3.5° N to M80.

M19 NGC 6273 GL. Cl. 17h 02.6m −26° 16′ mag: 7.4 size: 6′ dia.
From α Sco. go 0.1° N & 7.5° E to M19.

M62 NGC 6266 GL. Cl. 17h 01.2m −30° 07′ mag: 7.5 size: 5′ dia.
From M19 go 0.3° W & 3.9° S to M62.

Numbers 86 and 87   Stars to mag. 9.0

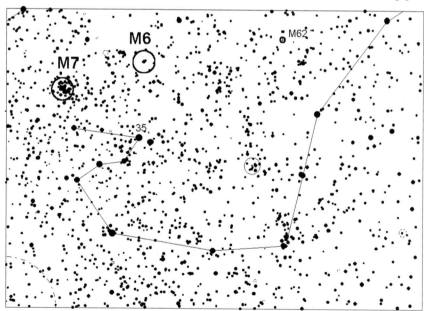

M6 NGC 6405 Open Cl. 17h40.0m −32° 15′ mag: 5.3 size: 30′ × 20′.
From M62 go 2.1° S & 8.4° E to M6.

M7 NGC 6475 Open Cl. 17h54.0m −34° 49′ mag: 5.0 size: 32′ × 26′.
From M6 go 2.6° S & 2.9° E to M7.

147

M11, M26, M16, M17 and M18

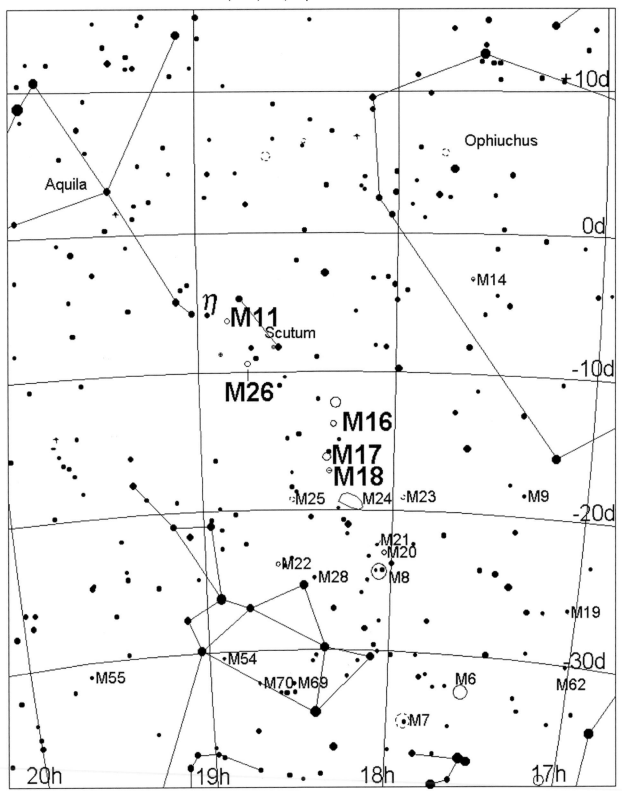

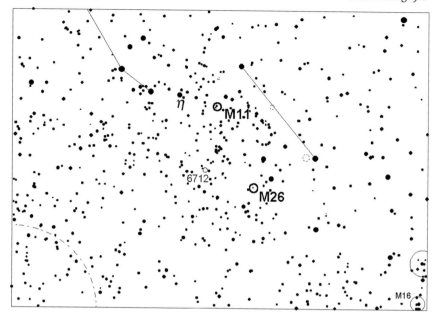

M11 NGC 6705 Open Cl. 18h51.1m −06° 16′ mag: 6.0 size: 8′ dia.
From η Sct. go 0.4° S & 1.4° W to M11.

M26 NGC 6694 Open Cl. 18h45.2m −09° 24′ mag: 7.8 size: 7′ dia.
From M11 go 1.5° W & 3.1° S to M26.

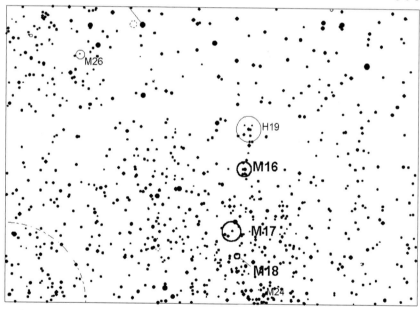

M16 NGC 6611 Open Cl. 18h 18.9m −13° 47′ mag: 6.8 size: 25′ dia.
From M26 go 4.3° S & 6.4° W to M16.

M17 NGC 6618 Diff. Neb. 18h21.1m −16° 10′ mag: 6.6 size: 15′ × 8′.
From M16 go 0.5° E & 2.4° S to M18.

M18 NGC 6613 Open Cl. 18h19.9m −17° 08′ mag: 7.1 size: 8′ × 6′.
From M17 go 0.2° W & 1.0° S to M18

M24, M23, M25, M21, M20, M8, M28 and M22

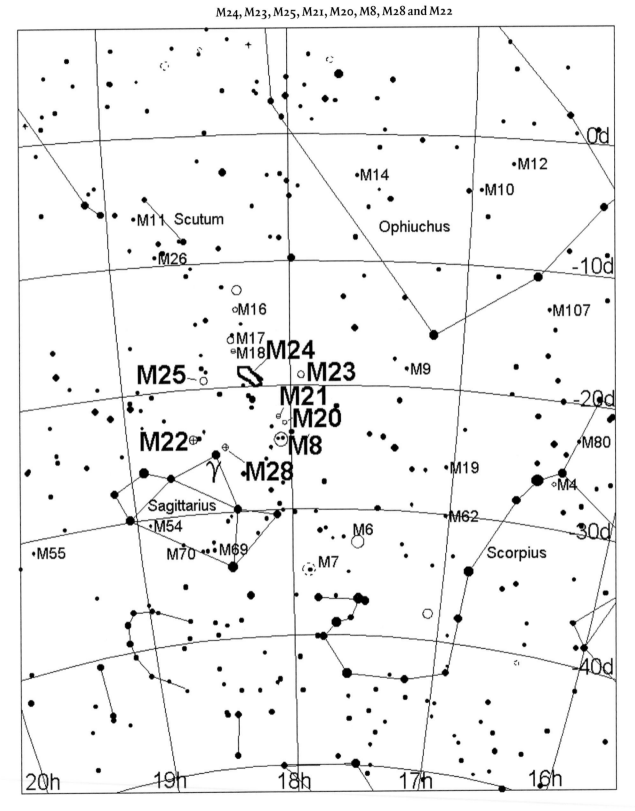

Numbers 93, 94, 95, 96, 97, 98, 99 and 100   Stars to mag. 9.5

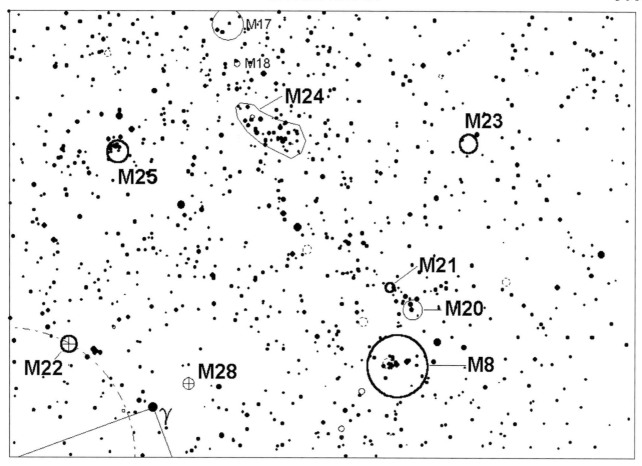

M24 Star Cloud 18h 18.4m −18° 25′ mag: 4.5 size: 80′ × 35′.
From M18 go 0.4° W & 1.2° E to M24.

M25 IC 4725 Open Cl. 18h31.7m −19° 07′ mag: 5.4 size: 35′ × 30′.
From M24 go 0.8° S & 3.1° E to M25.

M23 NGC 6494 Open Cl. 17h56.9m −19° 01′ mag: 6.2 size: 20′ × 15′.
From M25 go back to M24. Then go 0.6° S & 3.7° W to M23.

M21 NGC 6531 Open Cl. 18h04.7m −22° 30′ mag: 7.1 size: 7′ × 4′.
From M23 go 1.8° E & 3.5° S to M21.

M20 NGC 6514 Diff. Neb. 18h02.4m −23° 02′ mag: 6.8 size: 16′ × 10′.
From M21 go 0.5° W & 0.5° S to M20.

M8 NGC 6523 Diff. Neb. 18h03.7m −24° 23′ mag: 5.2 size: 25′ × 20′.
From M20 go 0.3° E & 1.4° S to M8.

M28 NGC 6626 GL. Cl. 18h24.6m −24° 52′ mag: 7.8 size: 5′ dia.
From M8 go 0.5° S & 4.8° E to M28.

M22 NGC 6656 GL. Cl. 18h36.4m −23° 54′ mag: 5.4 size: 17′ dia.
From M28 go 1.0° N & 2.7° E to M22.

M69, M70, M54, M55 and M75

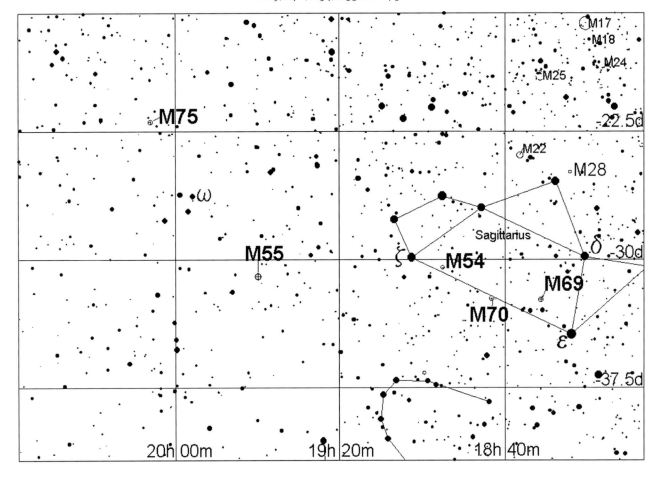

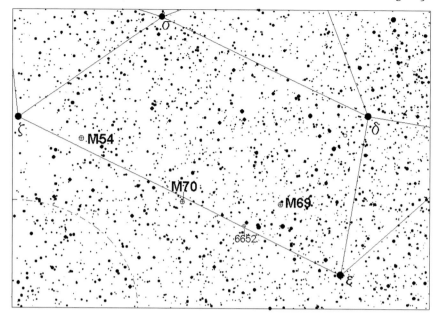

M69 NGC 6637 GL. Cl. 18h31.4m −32° 21′ mag: 7.9 size: 3′ dia.
From ε Sgr. go 1.4° E & 2.0° N to M69.

M70 NGC 6681 GL. Cl. 18h43.2m −32° 17′ mag: 8.4 size: 2′ dia.
From M69 go 2.5° E to M70.

M54 NGC 6715 GL. Cl. 18h55.1m −30° 28′ mag: 7.7 size: 3′ dia.
From M70 go 2.5° E & 1.8° N to M54.

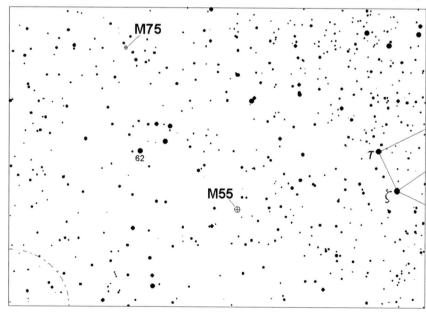

M55 NGC 6809 GL. Cl. 19h40.0m −30° 57′ mag: 6.1 size: 12′ dia.
From ζ Sgr. go 1.1° S & 8.0° E to M55.

M75 NGC 6864 GL. Cl. 20h06.1m −21° 55′ mag: 8.7 size: 2′dia.
From M55 go 3.2° N & 5.1° E to the star '62' Sgr. Then go 0.7° E & 5.8° N to M75.

M15, M2, M72, M73

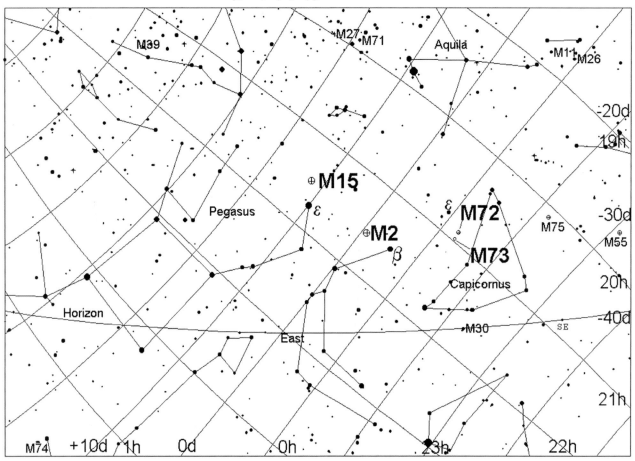

The eastern horizon before morning
twilight in late March from mid-northern
latitudes.

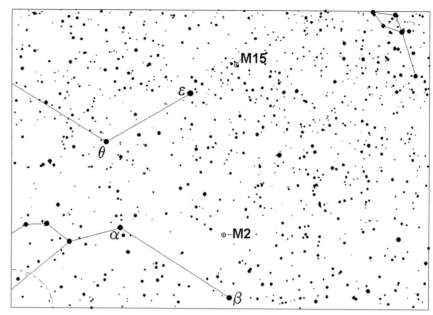

M15 NGC 7078 GL. Cl. 21h30.0m +12° 10′ mag: 6.8 size: 7′ dia.
From ε Peg. go 2.3° N & 3.5° W to M15.

M2 NGC 7089 GL. Cl. 21h33.5m −00° 49′ mag: 7.0 size: 8′ dia.
From M15 go 0.6° E & 13.0 S to M2; or from β Aqr. go 0.5° E & 4.7° N to M2.

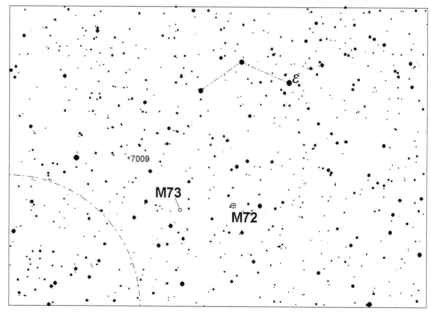

M72 NGC 6981 GL. Cl. 20h53.5m −12° 32′ mag: 9.0 size: 4′ dia.
From ε Aqr. go 1.2° E & 3.0° S to M72.

M73 NGC 6994 Open Cl. 20h59.0m −12° 38′ mag: 9.2 size: 2′ dia.
From M72 go 0.1° S & 1.3 E to M73.

M30

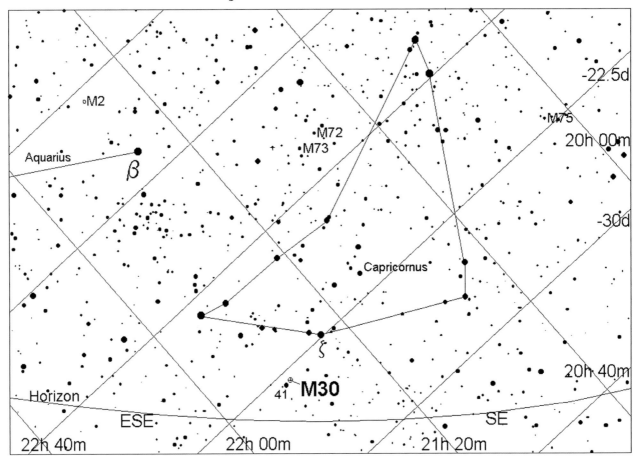

The eastern horizon before morning
twilight in late March from mid-northern
latitudes.

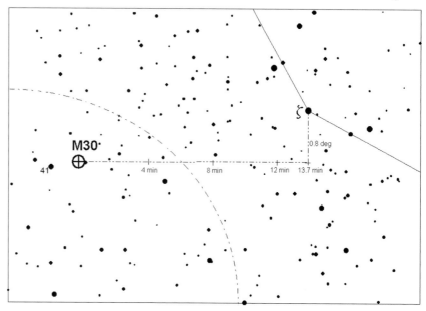

The time in minutes until M30 arrives.

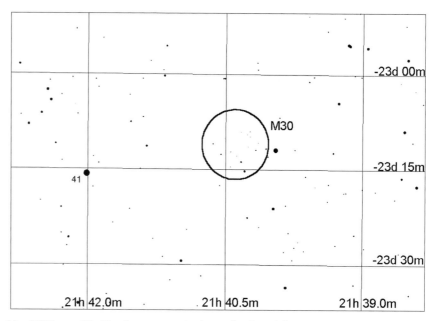

M30 NGC 7099 GL. Cl. 21h40.4m −23° 11′ mag: 8.1 size: 4′ dia.
From ζ Cap. go 0.8° S & 3.2° E to M30.

Good Morning!